Polymetallic Coatings to Control Biofouling in Pipelines

Polymetallic Coatings to Control Biofouling in Pipelines
Challenges and Potential

Vinita Vishwakarma

Dawn S S

K. Gobi Saravanan

A. M. Kamalan Kirubaharan

Saravanamuthu Vigneswaran

Gayathri Naidu

CRC Press
Taylor & Francis Group
Boca Raton London New York

CRC Press is an imprint of the
Taylor & Francis Group, an **informa** business

First edition published 2022
by CRC Press
6000 Broken Sound Parkway NW, Suite 300, Boca Raton, FL 33487-2742

and by CRC Press
2 Park Square, Milton Park, Abingdon, Oxon, OX14 4RN

© 2022 Vinita Vishwakarma, Dawn S S, K. Gobi Saravanan, A. M. Kamalan Kirubaharan, Saravanamuthu Vigneswaran, and Gayathri Naidu

CRC Press is an imprint of Taylor & Francis Group, LLC

ISBN: 9781032044897 (hbk)
ISBN: 9781032044903 (pbk)
ISBN: 9781003193449 (ebk)

Typeset in Palatino
by codeMantra

Contents

Contents

List of Tables

List of Figures

Preface

Biofouling is the major concern in most of the industries. Fouling-released coating with a special emphasis on pipeline industries is the focus of this book.

In oil and gas/fuel pipelines and storage tanks, biofilms cause significant operational problems due to microbial invasion, which leads to reduction of flow, souring, and reservoir plugging, thereby enhancing the corrosion of the bacterial adhered surface. With the constant use of biofuels, the study of the corrosive processes associated with microorganisms has also gained significance. Carbon steel (CS) or stainless steel is principally used for the transport of these materials because it is efficient and cost-effective. The major problems faced by these materials are the aggressive environment of fouling attack. Therefore, the study of protecting these pipeline surfaces with a protective coat is therefore quite significant. The surfaces of the pipeline colonize with the biofilm and form the complex microbial structure such as extracellular polymeric substance (EPS). The most abundant bacterial species, which are involved in the internal corrosion of the pipelines, interact with molecular hydrogen present on the surfaces of pipes and produce hydrogen sulfide as a by-product of metabolism. This process breaks down the iron and steel of even heavy-walled pipes, resulting in leaks and catastrophic pipeline failure, and thus reduces the quality of oil and gas/biofuel pipelines and storage tanks, therefore causing great financial losses worldwide. Surface modification of metallic materials by thin-film technology has achieved a considerable breakthrough to enhance its properties. Coatings on the surface provide a barrier between the surface and the environment, and provide fouling resistance to the surface. Various techniques have been used for surface modification such as chemical vapor deposition (CVD), physical vapor deposition (PVD), electrochemical and electroless plating. CVD and PVD techniques are relatively expensive and sophisticated. The choice for metallic coatings in recent years has emerged due to their low cost, fast deposition, good filling capability, good uniformity, and low processing temperature. These metallic coatings have been used extensively in the aerospace, automotive, and chemical processing industries in the past decade. It is the easiest method to deposit metallic films on arbitrary shapes with uniform thickness. It is a good choice of metal coating for oil and gas/biofuel pipelines and storage tanks, because of its excellent corrosion resistance and wear resistance when used as a barrier layer. It also has the benefit of an adjustable thickness of coating, good cohesion, and performance. Metal nanoparticles are known to exhibit enhanced physical and chemical properties when compared to their bulk counterparts because of their high surface-to-volume ratios. Copper (Cu), zinc (Zn), silver (Ag), nickel (Ni), etc. are known to have excellent toxicity for fouling organisms and thereby provide good resistance to biofouling. The fouling resistance is mostly due to toxic metallic ions on the surface, making it inhospitable to most organisms by blocking the respiratory enzyme system of these microorganisms in addition to damaging microbial DNA and the cell wall. Various surface characterization studies are there to evaluate the nanocrystalline nature of the film such as X-ray surface analysis (XRD), Field Emission Scanning Electron Microscopy (FESEM), Atomic Force Microscopy (AFM), thickness and contact angle measurement, corrosion behavior to distinguish the crystalline nature, morphology, chemical composition, and surface topography of the coating surface. The mechanism of corrosion resistance/improvement in the deposited layer should be studied by electrochemical techniques. Raman spectroscopy is used to characterize the corrosion deposits on the metal surface to find out the iron oxide phases and their transformation. Leaching of deposited materials study should be performed by using Inductively Coupled Plasma Mass Spectrometry (ICPMS)/Atomic Absorption Spectroscopy (AAS). The microbial culture methods such as isolation and identification of microbes are used to identify the biofilm and corrosion-causing bacteria, which are significant in diversity study and find the variation in the bacterial community.

The book contains eight chapters. The first part of the book discusses on biofouling mechanisms, assessment, and reduction strategies followed by biofouling in oil and gas/biofuel pipelines. The second part consists of surface coating methods to prevent the fouling and deterioration of the surface of pipes. The third part includes the ethical issues and environmental safety of coating and prospects of coating technologies.

Biofouling is the major concern in most of the industries. Fouling-released coating with special emphasis on pipeline industries is the better prospective for future direction research.

This book will be useful to learn the interdisciplinary skill for Chemical Engineering, Mechanical Engineering, and Biotechnology undergraduate students.

Vinita Vishwakarma
Dawn S S
K. Gobi Saravanan
A. M. Kamalan Kirubaharan
Saravanamuthu Vigneswaran
Gayathri Naidu

Authors

Dr. Vinita Vishwakarma is a Professor (Research) at the Centre for Nanoscience and Nanotechnology, Sathyabama Institute of Science and Technology, Chennai, India. She received her Ph.D. from Ranchi University (Ranchi) in 2002 and has since published more than 100 research articles, book chapters, and books in the fields of Biofouling, Biocorrosion, Thin films, Bioimplants, Surface modifications, Concrete corrosion, and Environment Health and Safety. Her research is focused on the development of metals and materials to control Biofouling and Biocorrosion. She has received sponsored research projects from various funding of the Government of India. She is a member of many professional bodies such as the Indian Science Congress Association, the Indian Institute of Metals, and the Indian Women Scientist Association and Organization for Women in Science for the Developing World. She has also edited several proceedings and books and is a reviewer in many journals.

Dr. Dawn S S is a Chemical Engineer and has over 20 years of experience in teaching and research. She is a Professor (Research) and Head in the Centre for Waste Management, a Centre of Excellence for Energy Research. She has made over 50 publications in reputed journals and conference proceedings. She has written a book entitled *Bio & Enzyme Engineering* and has authored several book chapters. She is a Consultant in Wasmanpro Solutions, Chennai and has experience in Environmental Impact Assessment. Waste management for energy and value-added products recovery and wastewater assessment and treatment are her interests. Biofuels from waste resources and their associated impacts on materials during transportation and storage and organic coatings on metal surfaces are also her focus. She has operated projects funded by the Indian Space Research Organization, Department of Science and Technology and Ministry of Human Resource Development, and Government of India. She is actively involved in rural women development through projects funded by Unnat Bharat Abhiyan by providing skill training for the women for their livelihood.

Dr. K. Gobi Saravanan is Scientist "C" at the Centre for Nanoscience and Nanotechnology, Sathyabama Institute of Science and Technology, Chennai. Currently, he is working on biomaterials, surface modification, antimicrobial coatings, corrosion protection of implant materials, and bioactive and biocompatibility materials. He has published 25 papers in reputed international journals and many papers in national/international conferences and has two Indian patents. In addition, he has attracted funding from various government funding agencies like the Department of Biotechnology, Indian Space Research Organization, etc.

Dr. A. M. Kamalan Kirubaharan is Scientist "C" at the Centre for Nanoscience and Nanotechnology, Sathyabama Institute of Science and Technology, Chennai. He has research experience in reputed organizations like ISRO, CSIR-CECRI, and IGCAR. His research expertise lies in the fields of high-temperature coatings, corrosion protection, geopolymer-based coatings, thin films, biosensors, and phase transformations of materials. He has authored 25 papers in reputed international journals and seven book chapters and has presented his work in many national/international conferences. He has filed six Indian patents. He has received seven research grants from various government funding agencies like the Indian Space Research Organization (ISRO) and Defense Research and Development Organization (DRDO). He is a reviewer for several reputed international journals. He is a recipient of IOP Trusted Reviewer Award 2021.

Prof. Saravanamuthu Vigneswaran is currently a Distinguished Professor of Environmental Engineering in the Faculty of Engineering and IT at the University of Technology, Sydney (UTS), Australia. He obtained his Doctor of Engineering and Dr.Sc. from the University of Montpellier and University of Toulouse, respectively. He is a Distinguished Fellow of the International Water Association. During the last 25 years, he has made significant contributions to the understanding of membrane systems in water reuse, resource recovery and desalination, and sustainable water systems in developing countries. He has several national and international projects in these areas. He has obtained over 15 Australian Research Council grants and worked as investigator and work package leader in four EU projects. He has served as UTS coordinator of several consortiums: Australian National Centre of Excellence for Desalination, Cooperative Research Centre for Contamination Assessment and Remediation of the Environment (CRC CARE). He has published over 350 journal papers, books, and book chapters.

He holds several international and national awards, including Google Impact Challenge Technology against Poverty Prize (2017), IWA Global Project Innovation and Development Awards in Research (2012,2019), and Kamal Fernando Mentor Award (2018).

Dr. Gayathri Naidu obtained her Ph.D. and postdoctoral research fellowship from the School of Civil and Environmental Engineering at the University of Technology Sydney (UTS). Her research interest is focused on water and wastewater treatment, membrane distillation, nanoparticles, seawater mining, and integrated resource recovery from acid mining water. She has worked as an industrial process engineer with Motorola and gained experience as a water and wastewater policy specialist. She is a member of the Membrane Society of Australasia (MSA). She has published more than 40 journal papers and one book chapter to her credits.

Introduction

Biofouling is the major concern in most of the industries, which is caused by different types of microorganisms including bacteria. These bacteria perform their metabolic activities on the metal surface. The types of bacteria involved in such activities are sulfate-reducing bacteria, iron-reducing bacteria, carbon dioxide-reducing bacteria, sulfur-reducing bacteria, and iron- and manganese-oxidizing bacteria. Among all these bacteria, sulfate-reducing bacteria are involved majorly in anaerobic corrosion. These microorganisms are responsible for the biodegradation of petroleum in natural environments (Harayama et al. 1999). Hydrocarbons and some other organic compounds present in the petroleum products are the excellent food source for microorganisms (Al-Abbas et al. 2012). Oil and gas pipelines are often subjected to deterioration because of the microbial activities on the surface of the pipeline materials (Alabbas and Mishra 2013). It has been reported that internal cracking of pipelines of almost 70%–95% is due to the localized corrosion caused by microorganisms, which is also termed as microbially influenced corrosion (MIC) (Reza 2008; Little and Lee 2007). Sulfate-reducing bacteria (SRB) are the main culprit to deteriorate the pipeline infrastructure as they generate sulfide (H_2S) using sulfate ions (SO_4). This biogenic sulfide causes serious health problems and environmental threats.

There are some conventional physical and chemical methods of cleaning of biofouling that are available to clean the foulants to some extent. Physical methods such as backwashing, sponge ball cleaning, turbulence, etc. are used to remove the deposition of the fouling especially in the pipelines and condenser tubes (Qin 2002). Chemicals such as surfactants, electrolytes, acids, and alkalis are used to remove the membrane fouling. Fouling-released coating with a special emphasis on pipeline industries is the better prospective to get rid of these problems (Cristiani and Perboni 2014). Surface coating onto the materials is the alternative way to improve the antifouling as well as durability of the materials. Before coating, pretreatment and surface activation are the required steps and then suitable coating techniques such as dip coating, spin coating, thermal spraying, painting, powder coating, electroplating, electroless, plasma spray coating, magnetron sputtering, physical and chemical evaporation, and pulsed laser deposition are implemented based on the application and industrial demand of the materials (Abdeen et al. 2019; Charbonnier et al. 2006). After the coating process, proper characterization of the materials is required to understand the properties of it. X-ray diffraction (XRD) analysis was performed to know the crystallinity of the materials, atomic force microscopy (AFM) was performed to analyze the surface topography and roughness, scanning electron microscopy (SEM) was performed to observe the sample morphology, X-ray photoelectron spectroscopy (XPS) was used to obtain the elemental and chemical bonding information, transmission electron microscopy (TEM) was used for understanding the internal microstructure of materials, and Raman spectroscopy was performed to understand the variations in chemical composition (Naito et al. 2018; Obrosov et al. 2017). Apart from these instruments, epifluorescence and a confocal laser scanning microscope (CLSM) were used to monitor the bacterial adherence on the material surface (Adair et al. 2000).

It has been observed that the coating materials are risky for the workers from a health and safety point of view. Industries are using inorganic and organic chemicals to prevent the materials from corrosion. The inorganic coatings are safer than organic coatings. The upcoming coating trends are the use of nanomaterials for coatings (Tri et al. 2019). The workers must take extra care precautions while using these nanomaterials as the high surface area of these materials is easily absorbed by the body compared to lower surface areas of the materials (Asmatulu, Zhang, and Asmatulu 2013). Therefore, from the safety viewpoint, eco-friendly methods are drawing attention to prepare metal nanoparticles for the coating process and this is an effective way to protect the workers and reduce the hazardous coating materials. Another important point is that the release of coating materials should be controlled and minimum to the environment (Azeem, KuShaari, and Man 2016). Industries should make sure of the occupational health and safety of the workers in their premises. It is the responsibility of the employer to provide a complete physical, mental, and socially safe working environment without any accident in the working place to the employee. Appropriate education and training are required for the workers to safeguard them from the ethical issues involved during the coating process. Coating industries are still looking forward to the new possibilities to protect the metals and materials from microbial corrosion to enhance their structural, functional, and esthetic properties.

There are various conventional methods to perform the coating on materials such as stainless steel (SS), titanium (Ti), aluminum (Al), carbon steel (CS), etc. The traditional solvent-borne coating, polymer coating, and solid and powder coatings are conventionally used for protecting the materials from chemical and microbial corrosion. Commercially, petrochemical-based polyols are used in polyurethane, whereas the price of crude petroleum is high and of environmental concern (Ghasemlou et al. 2019; Gama, Ferreira, and Barros-Timmons 2018). Researchers are finding that the vegetable oil-based polyols are anticorrosive, low toxic, thermally stable, and biodegradable unique triglyceride structures of low cost (Alagi, Choi, and Hong 2016; Alagi and Hong 2015; Maisonneuve et al. 2016; Zhang et al. 2014). But still, virgin vegetable oil is expensive, and it is an edible source to make polyurethane. So, to avoid the competition of resources, waste cooking oil (WCO) should be explored as a raw material to make polyurethane as it has similarities in the organic structure that make WCO suitable as a source to synthesize polyols using the same reaction as used to produce virgin vegetable oil-based polyols (Salleh, Tahir, and Mohamed 2018; Mohd Tahir et al. 2016; Lubis et al. 2019; Enderus and Tahir 2017). In India, not much work has been done on the WCO-based polyurethane coatings and their application on pipelines. To enhance the properties of WCO-based polyurethane coatings, nanoparticles suitable to enhance the coating properties must be incorporated.

The future innovative coating technology should not use hazardous materials for coatings, and we should consider the recycled materials as 'green coatings' by incorporating the waste materials to avoid harmful chemicals, reduce carbon footprints, release volatile organic compounds, and reduce the waste. The future of green coating is booming, as it is environmentally friendly and enhances the functionality to the surface.

References

Abdeen, Dana H., Mohamad El Hachach, Muammer Koc, and Muataz A. Atieh. 2019. "A review on the corrosion behaviour of nanocoatings on metallic substrates." *Materials* 12(2): 210. MDPI AG.

Adair, Colin G., Sean P. Gorman, Lisa B. Byers, David S. Jones, and Thomas A. Gardiner. 2000. "Confocal laser scanning microscopy for examination of microbial biofilms." In An, Y. H. and Friedman, R. J. (Ed.), *Handbook of Bacterial Adhesion*, 249–57. Totowa, NJ: Humana Press.

Al-Abbas, F. M., John Spear, Anthony Kakpovbia, N. M. Balhareth, David Olson, and Brajendra Mishra. 2012. "Bacterial attachment to metal substrate and its effects on microbiologically-influenced corrosion in transporting hydrocarbon pipelines." *The Journal of Pipeline Engineering* 11 (March): 63–72.

Alabbas, Faisal M., and Brajendra Mishra. 2013. "Microbiologically influenced corrosion in the gas industry." *In The 8th Pacific Rim International Congress on Advanced Materials and Processing*, 62: 3441–8.

Alagi, Prakash, and Sung Chul Hong. 2015. "Vegetable oil-based polyols for sustainable polyurethanes." *Macromolecular Research* 23 (12). Polymer Society of Korea: 1079–86.

Alagi, Prakash, Ye Jin Choi, and Sung Chul Hong. 2016. "Preparation of vegetable oil-based polyols with controlled hydroxyl functionalities for thermoplastic polyurethane." *European Polymer Journal* 78 (May). Elsevier Ltd.: 46–60.

Asmatulu, Ramazan, B. Zhang, and E. Asmatulu. 2013. "Safety and ethics of nanotechnology." In Asmatulu, Ramazan (Ed.), *Nanotechnology Safety*, 31–41. Fairmount, Wichita, KS, Elsevier B.V.

Azeem, Babar, Kuzilati KuShaari, and Zakaria Man. 2016. "Effect of coating thickness on release characteristics of controlled release urea produced in fluidized bed using waterborne starch biopolymer as coating material." *Procedia Engineering*, 148:282–89. Elsevier Ltd.

Charbonnier, M., Maurice Romand, Y. Goepfert, D. Léonard, and M. Bouadi. 2006. "Copper metallization of polymers by a palladium-free electroless process." *Surface and Coatings Technology* 200 (18–19). Elsevier: 5478–86.

Cristiani, P., and G. Perboni. 2014. "Antifouling strategies and corrosion control in cooling circuits." *Bioelectrochemistry* 97 (June). Elsevier: 120–26.

Enderus, N. F., and S. M. Tahir. 2017. "Green waste cooking oil-based rigid polyurethane foam." *IOP Conference Series: Materials Science and Engineering* 271: 012062. Institute of Physics Publishing.

Gama, Nuno, Artur Ferreira, and Ana Barros-Timmons. 2018. "Polyurethane foams: Past, present, and future." *Materials* 11 (10). MDPI AG: 1841.

Ghasemlou, Mehran, Fugen Daver, Elena P. Ivanova, and Benu Adhikari. 2019. "Polyurethanes from seed oil-based polyols: A review of synthesis, mechanical and thermal properties." *Industrial Crops and Products* 142: 111841. Elsevier B.V.

Harayama, S, H. Kishira, Y. Kasai, and K. Shutsubo. 1999. "Petroleum biodegradation in marine environments." *Journal of Molecular Microbiology and Biotechnology* 1 (1). Switzerland: 63–70.

Little, Brenda J., and Jason S. Lee. 2007. *Microbiologically Influenced Corrosion*. Hoboken, NJ: Wiley & Sons.

Lubis, Maulida, Mara Bangun Harahap, Iriany Iriany, Muhammad Hendra S. Ginting, Iqbal Navissyah Lazuardi, and Muhammad Amri Prayogo. 2019. "The utilization of Waste Cooking Oil (WCO) as a mixture of polyol sourcein. The production of polyurethane using toluene diisocyanate." *Oriental Journal of Chemistry* 35 (1). Oriental Scientific Publishing Company: 221–27.

Maisonneuve, Lise, Guillaume Chollet, Etienne Grau, and Henri Cramail. 2016. "Vegetable oils: A source of polyols for polyurethane materials." *OCL - Oilseeds and Fats, Crops and Lipids*. EDP Sciences.

Mohd Tahir, Syuhada, Wan Norfirdaus Wan Salleh, Nur Syahamatun Nor Hadid, Nor Fatihah Enderus, and Nurul Aina Ismail. 2016. "Synthesis of waste cooking oil-based polyol using sugarcane bagasse activated carbon and transesterification reaction for rigid polyurethane foam." *Materials Science Forum*, 846:690–96.

Naito, Makio, Toyokazu Yokoyama, Kouhei Hosokawa, and Kiyoshi B. T. 2018. "Chapter 5- Characterization Methods for Nanostructure of Materials." In Naito, Makio, Yokoyama, Toyokazu, Hosokawa, Kouhei, Nogi, Kiyoshi (Ed.), *Nanoparticle Technology Handbook* (Third Edition), 255–300. Osaka: Elsevier.

Obrosov, Aleksei, Roman Gulyaev, Andrzej Zak, Markus Ratzke, Muhammad Naveed, Wlodzimierz Dudzinski, and Sabine Weiß. 2017. "Chemical and morphological characterization of magnetron sputtered at different bias voltages Cr-Al-C coatings." *Materials* 10 (2): 156. MDPI AG.

Qin, Jianying. 2002. "Sponge-ball automatic tube cleaning device for saving energy in a chiller." *International Energy Journal* 3 (June): 35–43.

Reza, Javaherdashti. 2008. *Microbiologically Influenced Corrosion (MIC) an Engineering Insight. Microbiologically Influenced Corrosion*. London: Springer London.

Salleh, W. N. F. W., S. M. Tahir, and N. S. Mohamed. 2018. "Synthesis of waste cooking oil-based polyurethane for solid polymer electrolyte." *Polymer Bulletin* 75 (1). Springer Verlag: 109–20.

Tri, Phuong Nguyen, Tuan Anh Nguyen, Sami Rtimi, and Claudiane M.Ouellet Plamondon. 2019. "Nanomaterials-based coatings: An introduction." In *Nanomaterials-Based Coatings: Fundamentals and Applications*, 1–7. Elsevier.

Zhang, Chaoqun, Yuzhan Li, Ruqi Chen, and Michael R. Kessler. 2014. "Polyurethanes from solvent-free vegetable oil-based polyols." *ACS Sustainable Chemistry and Engineering* 2(10): 2465–76. American Chemical Society.

1

Biofouling and Biocorrosion

Vinita Vishwakarma and K. Gobi Saravanan

CONTENTS

1.1 Introduction

The root cause of biofouling is the formation of a bacterial biofilm, where the growth of bacteria, fungi, algae, protozoans, and crustaceans leads to an undesirable accumulation of living organisms on the surfaces of metals and materials (Malik et al. 2012). Their secretions are the cause of contamination and corrosion of these surfaces and decrease their efficiency (Choudhary and Schmidt-Dannert 2010). In this process, the network of microorganisms interact with each other and form multicellular communities (Davey and O'toole 2000). The uncontrollable growth of biofouling leads to the biocorrosion (microbial corrosion/microbially influenced corrosion – MIC) problem. The metabolic activity of microorganisms accelerates the corrosion kinetics rate of a metallic substrate (Lenhart et al. 2014). Almost all the industries are affected by biofouling and biocorrosion problems and face huge economic loss every year. The marine components and their corrosion and deterioration are always associated with the biofouling mechanism, where corrosion mediated by sulfate-reducing bacteria is most common and plays a crucial role in biofouling, biocorrosion, and biodeterioration (Beech et al. 2006).

1.2 Microbial Growth and Biofouling

Microbial growth is the growth of a microorganism by number and not by size under the optimum temperature, pH, osmotic pressure, chemical requirements, nutrients, and so on. There are four growth phases of microbial growth (Figure 1.1):

    i. **Lag phase:** This phase is the metabolic activity phase of a microorganism in which an individual organism grows.

    ii. **Log phase:** This is the highest phase of metabolic activity of a microorganism.

    iii. **Stationary phase:** This phase is where the cell growth of a microorganism slows down and stabilizes.

    iv. **Death phase:** This phase occurs when the population size of a microorganism begins to decrease.

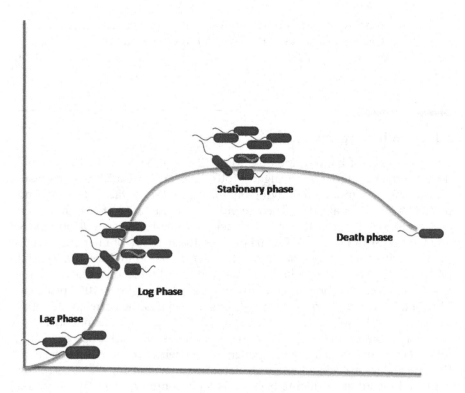

FIGURE 1.1
Growth phases of microbial growth.

1.3 Measurements of Microbial Growth

There are two methods to measure the microbial growth: direct and indirect methods.

1.3.1 Direct Methods of Measurements

The most common practice to measure viable cells is the plate count method, where a Petri dish must be inoculated with a sample and the colonies are counted. Pour plating is one of the plate count methods where the inoculum must be introduced into an empty Petri dish and then a liquid nutrient medium is added, followed by solidification and incubation at 37°C. Spread plating is another method to examine the microbial growth where the inoculum must be spread on the surface of the solid agar medium in a Petri dish. Another direct method to measure the microbial growth is filtration, which is used to measure small quantities of bacteria. The water samples must be filtered to retain the bacteria, which have to be transferred to a Petri dish and further incubated for counting the colonies. The most probable number is another method, where the sample must be inoculated in broth tubes after several dilutions and the probability of population is determined by a statistical method. In direct microscopic count, bacterial suspension (0.01 mL) will be placed on a microscopic slide and stained to visualize the bacteria.

1.3.2 Indirect Methods of Measurements

Turbidity, metabolic activity, and dry weight are the indirect methods for the measurement of microbial growth. The turbid sample of the bacteria will be used to determine the % transmission or absorbance using a spectrophotometer. The bacteria multiply in the liquid media and produce carbon dioxide and acids, which are measured to determine the metabolic activity of bacteria. To determine the dry weight, the liquid media must be centrifuged, and the cell pellet is weighed.

1.4 Biofouling

Biofouling is a serious problem all over the world. The growth of biofilms is referred to as biofouling, which is the accumulation of micro- and macroorganisms on a solid substrate. Certain environmental factors like temperature, pH, light, turbidity, etc. are responsible for the development of biofouling. This is happening extensively in many systems and processes such as medical industries equipment, different types of pipelines of water, oil and gas, food, dairy and wine industries, nuclear power plants, ship hulls, concrete structures, and water membrane system. This fouling attachment destroys the process of the industries, leading to operational failure, reduction of heat transfer efficiency, chemical and biocorrosion, and finally, to shut down.

Chemical fouling has a great impact on the heat exchangers as it decreases the thermal and mechanical performance, and thus, it enhances the corrosion rate in the pipes and lowers the heat transfer coefficient. Fouling problems in the pipelines are because of chemicals, precipitation, crystallization, particulates, corrosion, solidification, and biological microorganisms. Based on this, fouling can be classified into the following types:

1.5 Chemical Fouling

Chemical fouling occurs because of chemical changes in a fluid, which lead to scaling formation and significantly affect the performance of the equipment.

Precipitation fouling: The initiation of precipitation fouling is because of different types of salt solubility at different temperatures. Mostly calcium and magnesium carbonate and sulfate and silica adheres on the surface of the materials.

Crystallization fouling: Crystallization fouling or scaling is formed on the surface under the boiling condition which is common in heat exchanger surface (Bott 1997).

Particulate fouling: The particulate fouling is happening because of interactions between particles and fluid, and then its interaction with the surface. These deposits grow further and block the components by agglomeration and clogging (Henry, Minier, and Lefèvre 2012).

Corrosion fouling: The corrosion fouling is because of the metal oxides on the metal substrate such as iron oxide deposits on carbon steel.

Solidification fouling: Freezing of flowing fluid is solidification or freezing fouling which depends on the fluid velocity and temperature, for example, freezing of paraffin from a cooled petroleum product.

1.6 Biological Fouling

Biological fouling is due to the accumulation of microorganisms on the surface (Figure 1.2).

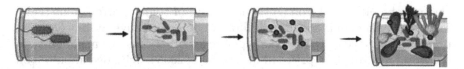

FIGURE 1.2
Biological fouling is due to the accumulation of microorganisms.

1.6.1 Biocorrosion and Biodeterioration

The process in which microorganisms corrode metals and metal alloys is called **biocorrosion**. Biocorrosion is also termed as Microbially Influenced Corrosion (MIC). Both aerobic and anaerobic bacteria participate in this process to corrode materials. The microorganisms perform metabolic activities, and the products of those activities are responsible for the deterioration of the materials. These include enzymes, acids (both organic and inorganic), volatile compounds such as ammonia or hydrogen sulfide, exopolymers, etc., which change the electrochemistry at the metal surface by cathodic and anodic reactions (Beech and Gaylarde 1999). In addition, the surrounding environment's pH, conductivity, oxygen and nutrient concentration, salinity, pressure, and redox potentials influence the properties of materials, which leads to corrosion. The sulfate-, iron-, and manganese-oxidizing bacteria are accountable for the corrosion under aerobic conditions. The sulfate-oxidizing bacteria such as *Thiobacillus* oxidize sulfur to sulfuric acid and are liable for oxidative corrosion of the materials. The iron-oxidizing bacteria (IOB) are mostly prone to industrial components. The IOB such as *Acidithiobacillus ferrooxidans* change the pH of the solution and develop a thick layer of Fe_2O_3 on the surface of stainless steel, which triggers corrosion (Wang et al. 2014). This corrosion product enhances the growth of sulfate-reducing bacteria (SRB), and the combined effect of these two bacteria is remarkably high (Xu et al. 2007).

Another IOB such as *Sphaerotilus* sp. is responsible for pitting corrosion in a 3% NaCl solution (Starosvetsky et al. 2008). Manganese-oxidizing bacteria consist of aerobic bacteria such as *Bacillus, Pseudomonas, Staphylococcus, Micrococcus*, and so on. These bacteria are responsible for corrosion failure in stainless steel, cast iron, mild steel, etc. Biocorrosion is also a serious issue in the oil and gas pipelines, which is a risk during transportation, water utilities system, and power plant cooling system (Gu 2012; Rajasekar et al. 2010) as their wall is water wetted which stimulates the development of microorganisms inside the pipelines, which leads to pipe cracks and failure. SRB are the most common bacteria responsible for biocorrosion under anaerobic conditions. Generally, acidic degradation, electron movement, metal depolarization, polymerization, biofilm formation, and mineral formation mechanisms are involved in the biocorrosion process.

At this point, in oil pipelines, oxygen is removed by an oxygen scavenger or aerobic biofilm. Through the anaerobic iron corrosion, there are loss of electrons by elemental iron (Fe^0) and release of soluble ferrous ions (Fe^{2+}). The US government is spending billions of dollars to solve the biocorrosion problems every year. The Alaska pipeline leak that caused a major spike in the world regarding oil prices in the spring of 2006 heightened attention on biocorrosion. Biocorrosion is a threat to most of the industries as they operate their plants with water, which consists of microorganisms. It affects the day-to-day process of industries, decreases the product quality, and results in great economic loss. Most of the industries are not interested to spend their time and money to get rid of this problem as they are not aware of this threat.

There is a huge economic loss in the industrial sector because of biocorrosion. Medical, water, food, dairy, wine, marine, power plants, and oil and gas industries are often exposed to biofouling, which leads to pipe blockages, waterlogging, and reduction of the heat-transfer efficiency, hampering the operating system of plants (Vishwakarma 2020). Several times, these industries are unaware of the threat of the problems of biocorrosion in their components.

Biodeterioration is the adverse changes in the material properties because of vital activities of microorganisms. The original properties of materials will change due to microbial activities. There are two different types of biodeterioration – physical and chemical (Tucker 2015).

Physical biodeterioration: Some materials break down because of the growth and metabolic activities of microorganisms (fungi) and cause physical or mechanical biodeterioration. This type of fouling also reduces the appearance of the materials.

Chemical biodeterioration: Because of the metabolic activities of microorganisms, they use materials as food for energy sources, for example, deterioration of bread by fungi (*Aspergillus* and *Fusarium*). This is assimilatory biodeterioration. On the other side, there is a chemical change in the food because the metabolic activity of the microorganism and the pH of the food will vary, which results in the acidic condition of the food. This is called dissimilatory biodeterioration.

Different types of materials such as metals (Berk et al. 2001) and non-metals including teeth (de Mele and Cortizo 2000), textile (Buschle-Diller et al. 1994), leather (Orlita 2004), ceramic (Sand and Bock 1991), glass (Schabereiter-Gurtner et al. 2001), wood (Björdal, Nilsson, and Daniel 1999), food, paper (Fabbri et al. 1997), dairy products, concrete structures, paintings (Rubio and Bolívar 1997), rocks and stones, and pipelines are deteriorated because of microorganisms such as bacteria, algae, fungi, molluscs, barnacles, sponge, and so on. Dental plaque in the form of biofouling on the tooth surface leads to many oral diseases (Zhou, Wong, and Li 2020). During textile processing, almost all the stages of microbial growth affect and deteriorate the color, strength, and appearance of textile materials (Szostak-Kotowa 2004). In the leather industry, biodeterioration of leather is an aggressive activity by microorganisms during the processing and storage because of the amount of proteins, lipids, and carbohydrates, which is a suitable environment for the growth of bacteria (Falkiewicz-Dulík 2015).

Ceramic materials are frequently affected by biodeterioration that leads to physical, chemical, and esthetical damages of the materials (Coutinho et al. 2020; Coutinho, Miller, and Macedo 2015), which can be prevented by protective coatings of nanoparticles. Wood degrades because of bacteria, fungi, and insects, and that can be protected during the processing of wood. There are many complex reactions of microorganisms in food deterioration, which affect the loss of taste, color, nutritional value, and food safety. Concrete biodeterioration because of biogenic acid produced by acid-producing bacteria reduces the long-term performance of the infrastructure (Wei et al. 2013). The oil and gas industries' pipeline system has localized biodeterioration problems because the hydrocarbons and other organic compounds present are an exceptional food source for the microorganisms (Alabbas and Mishra 2013). The problems associated with biodeterioration in different materials and microorganisms responsible for this are mentioned in Table 1.1.

TABLE 1.1

Microorganisms Responsible for the Biodeterioration of Materials

Materials and Industrial Components	Problems Associated with Biofilms	Microorganisms Responsible for the Biodeterioration of Materials
Metals	Biocorrosion, cracks, reduced strength	
Medical implants	Susceptibility on the surface of the implants	*Staphylococcus aureus, Klebsiella pneumonia, Enterobacter, Candida*
Teeth	Periapical infection, root canal infection, tooth decay, caries	*Streptococcus mutans, Propionibacterium propionicum, Porphyromonas gingivalis, Porphyromonas endodontalis, Fusobacterium nucleatum, Prevotella orali*
Textile	Cellulose degradation	*Cellulomonas* sp., *Cellvibrio* sp., *Bacillus megaterium, Aspergillus niger, Tricoderma, myrothecium, Chaetominum*
Leather	Deterioration	*Clostridium histolyticum, Clostridium capitovale, Penicillium* sp., *Aspergillus* sp.
Pipelines	Corrosion, scaling, fouling	*Pseudomonas* sp., *Desulfovibrio vulgaris*, SRB
Wood	Deterioration	*E. coli, Clostridium xylanolyticum*
Dairy products	Leftover milk in equipment	*Pseudomonas fluorescens*
Breweries	Found in biofilms in tanks, filters, pasteurizers, and fillers	*Lactobacillus, Pediococcus, Pseudomonas, Bacillus, Candida, Debaryomyces*
Wine	Unpleasant smell, flavors	*Lactobacillus*

(Continued)

TABLE 1.1 *(Continued)*

Microorganisms Responsible for the Biodeterioration of Materials

Materials and Industrial Components	Problems Associated with Biofilms	Microorganisms Responsible for the Biodeterioration of Materials
Concrete structures	Deterioration of concrete	*Thiobacillus intermedius, Acidithiobacillus thiooxidans, Thiomonas perometablis, Fusarium* sp., *Penicillium oxalicum, Coniosporium uncinatum*
Rocks and stones	Deterioration, fouling	*Thiobacillus* spp., *Nitrosomonas* spp., *Flavobacterius, Pseudomonas*
Marine structure	Corrosion: degradation of the material, reduced performance	*Pseudomonas* sp., *Bacillus* sp., sulfate-reducing bacteria
Power plants	Corrosion: degradation of the material, reduced performance	*Pseudomonas* sp., *Desulfovibrio vulgaris*, sulfate-reducing bacteria
Oil and gas industry	Biocorrosion, pit formation	*Pseudomonas* sp., *Desulfovibrio vulgaris, Desulfotomaculum* sp.

1.6.2 Causes and Effects – Health and Environmental Perspectives

Biofouling damages the infrastructures of materials and destroys their physical and mechanical structures. Its contaminations depend on their surrounding environments and are dangerous for human health. Health effects associated with biofilms are frequently affecting human beings and devices of various industries. Biofilms on medical devices such as contact lens, catheters, implants, and moreover teeth, wound infection, burn and cuts, skin, and so on are easily infected and persistent for a long time. Biofouling spoils food, dairy products, and water and is a serious menace for public wellness and persists in the surroundings (Tasneem et al. 2018). Biofilm formation in medical devices such as implants, catheters, surgical tools, and other devices is a significant issue. It is difficult to eradicate a biofilm from this inert surface; however, a novel approach to prevent the biofilm infections of medical devices is the surface modification (Srivastava and Bhargava 2016). Food products, fruits, vegetables, meat, dairy (butter and cheese), and wine products are deteriorated by *Bacillus, Pseudomonas, Botrytis, Campylobacter, Clostridium*, etc. during storage and therefore lose their quality, which leads to serious health problems in humans (Alenezi 2016). To preserve these food products, edible nanoencapsulation, polymer coating, can reduce food contamination and enhance the shelf life (Aloui and Khwaldia 2016). The textile materials are deteriorated because of the breakdown of cellulose by microorganism enzymes that change the color of fabrics and lead to deterioration of fabrics, which is a concern for infections and allergies in humans (Boryo 2013).

The problem of biodeterioration is mostly overlooked by the construction industries that lead to structural failures especially in concrete structures, which are exposed to harsh environments. The surface pH of the concrete structure will reduce due to the atmospheric carbon dioxide and the action of microbial growth leading to the deterioration of concrete (Uthaman et al. 2019). The surface modification of the concrete structures with antimicrobial coatings and incorporation of nanoparticles enhance their strength and durability, reduce their porosity, and improve their resistance to sulfate attack, chloride permeability, calcium leaching, etc. (Uthaman et al. 2018). This will avoid health and safety issues in concrete industries. Fouling of oil and gas/biofuel pipelines is the major issue in petrochemical plants, where SRB corrode carbon steel pipelines and produce sulfide. This biogenic sulfide affects public health and the surrounding environments (Khouzani et al. 2019). Due to the lack of awareness about the causes and side-effects of biofouling, biocorrosion, and biodeterioration, the degradation rates of various materials and industrial components are increasing, which results in health problems of the community.

1.7 Case Studies

Case Study 1: Alaska Pipeline Leak

The Alaska pipeline leak intensified the significance of biofouling in 2006, where the massive oil spill hit Alaska's North Slope region that caused a major spike in the world. It was estimated that around 267,000 gallons of crude oil seeped from the pipeline because of biocorrosion. This incident had drawn the attention of biofouling and biocorrosion.

Case Study 2: Failure of Pipeline

In 1994, there was a failure of the underground steel pipeline transporting sour crude oil in Saudi Arabia within 3 years of installations. The failure is due to the microbiologically influenced corrosion of the lower surface of the pipeline 25.5 km long and 28 inch in diameter. Low flow velocity and an excessive amount of water were the main reasons of the failure of this pipeline, where the bacterial activities were rigorous. Moreover, it was found that sulfate-reducing bacteria were the main culprit in this internal pipeline corrosion.

Case Study 3: Crude Oil Spillage

In April 2017, a downstream pipeline of 864 mm outer diameter of the Glenavon pump station near Saskatchewan (Province in Canada) ruptured and approximately 990 cc of crude oil spillage occurred on farmland.

The failure analysis made upon loss of 78 cc of the crude oil aided the investigators to draw corrosion cracking to be the root cause of the pipeline leak. However, no injuries were evidently recorded.

Case Study 4: Pipeline of Peace River Mainline

In 1968, a pipeline of Peace River Mainline, TransCanada broke 6 times and leaked 17 times until 2014 due to corrosion.

Case Study 5: Stress Corrosion

In March 1965, a gas transmission pipeline of 32 inch in the north of Natchitoches, Louisiana burst because of stress corrosion. This pipeline also had a blast in May 1955 because of 930 feet breakdown.

References

Alabbas, Faisal M., and Brajendra Mishra. 2013. "Microbiologically influenced corrosion of pipelines in the oil & gas industry." In *Proceedings of the 8th Pacific Rim International Congress on Advanced Materials and Processing*, 3441–48. Springer International Publishing, Cham.

Alenezi, Amal. 2016. "Antimicrobial and antioxidative properties of berry extracts." *Journal of Food Processing & Technology* 07 (09):68.

Aloui, Hajer, and Khaoula Khwaldia. 2016. "Natural antimicrobial edible coatings for microbial safety and food quality enhancement." *Comprehensive Reviews in Food Science and Food Safety* 15 (6). Blackwell Publishing Inc.: 1080–103.

Beech, I. B., and C. C. Gaylarde. 1999. "Recent advances in the study of biocorrosion - an overview." *Revista de Microbiologia* 30(3): 177–190. Sociedade Brasileira de Microbiologia.

Beech, I. B., J. A. Sunner, C. R. Arciola, and P. Cristiani. 2006. "Microbially-influenced corrosion: Damage to prostheses, delight for bacteria." *International Journal of Artificial Organs* 29 (4). Wichtig Editore s.r.l.: 443–52.

Berk, Sharon G., Ralph Mitchell, Ronald J. Bobbie, Janet S. Nickels, and David C. White. 2001. "Microfouling on metal surfaces exposed to seawater." *International Biodeterioration and Biodegradation* 48. Elsevier: 167–75.

Björdal, C. G., T. Nilsson, and G. Daniel. 1999. "Microbial decay of waterlogged archaeological wood found in Sweden. Applicable to archaeology and conservation." *International Biodeterioration and Biodegradation* 43 (1–2). Elsevier Science Ltd: 63–73.

Boryo, D. E. A. 2013. "The effect of microbes on textile material: A review on the way-out so far." *The International Journal of Engineering and Science* 2 (8): 9–13.

Bott, T. R. 1997. "Aspects of crystallization fouling." *Experimental Thermal and Fluid Science* 14 (4). Elsevier Inc.: 356–60.

Buschle-Diller, G., S. H. Zeronian, N. Pan, and M. Y. Yoon. 1994. "Enzymatic hydrolysis of cotton, linen, ramie, and viscose rayon fabrics." *Textile Research Journal* 64 (5). Sage Publications: 270–79.

Choudhary, Swati, and Claudia Schmidt-Dannert. 2010. "Applications of quorum sensing in biotechnology." *Applied Microbiology and Biotechnology* 86(5): 1267–1279. Appl Microbiol Biotechnol.

Coutinho, Mathilda L., Ana Z. Miller, and Maria F. Macedo. 2015. "Biological colonization and biodeterioration of architectural ceramic materials: An overview." *Journal of Cultural Heritage*. Elsevier Masson SAS.

Coutinho, Mathilda L., João Pedro Veiga, Maria Filomena Macedo, and Ana Zélia Miller. 2020. "Testing the feasibility of titanium dioxide sol-gel coatings on Portuguese glazed tiles to prevent biological colonization." *Coatings* 10 (12): 1–20.

Davey, Mary Ellen, and George A. O'toole. 2000. "Microbial biofilms: From ecology to molecular genetics." *Microbiology and Molecular Biology Reviews* 64 (4). American Society for Microbiology: 847–67.

de Mele, M. Fernández Lorenzo, and M. C. Cortizo. 2000. "Biodeterioration of dental materials: Influence of bacterial adherence." *Biofouling* 14 (4). Harwood Academic Publishers GmbH: 305–16.

Fabbri, A. A., A. Ricelli, S. Brasini, and C. Fanelli. 1997. "Effect of different antifungals on the control of paper biodeterioration caused by fungi." *International Biodeterioration and Biodegradation* 39 (1). Elsevier: 61–5.

Falkiewicz-Dulík, Michalina. 2015. "Leather and leather products." In *Handbook of Material Biodegradation, Biodeterioration, and Biostablization*, 133–256. Elsevier, ChemTec Publishing, Totanto, Canada.

Gu, Tingyue. 2012. "New understandings of biocorrosion mechanisms and their classifications." *Journal of Microbial and Biochemical Technology* 4 (4): 3–6.

Henry, Christophe, Jean Pierre Minier, and Grégory Lefèvre. 2012. "Towards a description of particulate fouling: From single particle deposition to clogging." *Advances in Colloid and Interface Science* 185–186:34–76. Elsevier B.V.

Khouzani, Mahdi Kiani, Abbas Bahrami, Afrouzossadat Hosseini-Abari, Meysam Khandouzi, and Peyman Taheri. 2019. "Microbiologically influenced corrosion of a pipeline in a petrochemical plant." *Metals* 9 (4): 9040459.

Lenhart, Tiffany R., Kathleen E. Duncan, Iwona B. Beech, Jan A. Sunner, Whitney Smith, Vincent Bonifay, Bernadette Biri, and Joseph M. Suflita. 2014. "Identification and characterization of microbial biofilm communities associated with corroded oil pipeline surfaces." *Biofouling* 30 (7). Taylor and Francis Ltd.: 823–35.

Malik, Abdul, Mashihur Rahman, Mohd Ikram Ansari, Farhana Masood, and Elisabeth Grohmann. 2012. "Environmental protection strategies: An overview." In *Environmental Protection Strategies for Sustainable Development*, 1–34, Editor: Abdul Malik and Elisabeth Grohmann. Netherlands: Springer.

Orlita, Alois. 2004. "Microbial biodeterioration of leather and its control: A review." *International Biodeterioration and Biodegradation* 53. Elsevier: 157–63.

Rajasekar, Aruliah, Balakrishnan Anandkumar, Sundaram Maruthamuthu, Yen Peng Ting, and Pattanathu K. S. M. Rahman. 2010. "Characterization of corrosive bacterial consortia isolated from petroleum-product-transporting pipelines." *Applied Microbiology and Biotechnology* 85 (4). Appl Microbiol Biotechnol: 1175–88.

Rubio, R. F., and F. C. Bolívar. 1997. "Preliminary study on the biodeterioration of canvas paintings from the seventeenth century by *microchiroptera*." *International Biodeterioration and Biodegradation* 40. Elsevier Sci Ltd.: 161–69.

Sand, W., and E. Bock. 1991. "Biodeterioration of ceramic materials by biogenic acids." *International Biodeterioration* 27 (2). Elsevier: 175–83.

Schabereiter-Gurtner, Claudia, Guadalupe Piñar, Werner Lubitz, and Sabine Rölleke. 2001. "Analysis of fungal communities on historical church window glass by denaturing gradient gel electrophoresis and phylogenetic 18S RDNA sequence analysis." *Journal of Microbiological Methods* 47 (3). Elsevier: 345–54.

Srivastava, Shilpi, and Atul Bhargava. 2016. "Biofilms and human health." *Biotechnology Letters* 38(1): 1–22. Springer Netherlands.

Starosvetsky, J., D. Starosvetsky, B. Pokroy, T. Hilel, and R. Armon. 2008. "Electrochemical behaviour of stainless steels in media containing Iron-Oxidizing Bacteria (IOB) by corrosion process modeling." *Corrosion Science* 50 (2). Pergamon: 540–47.

Szostak-Kotowa, Jadwiga. 2004. "Biodeterioration of textiles." *International Biodeterioration and Biodegradation* 53. Elsevier: 165–70.

Tasneem, Umber, Nusrat Yasin, Iqbal Nisa, Faisal Shah, Ubaid Rasheed, Faiza Momin, Sadir Zaman, and Muhammad Qasim. 2018. "Biofilm producing bacteria: A serious threat to public health in developing countries." *Journal of Food Science and Nutrition* 01 (02): 25–31. Allied Academies.

Tucker, Gary S. 2015. "Control of biodeterioration in food." In *Food Preservation and Biodeterioration*, 1–35, Editor: Gary S. Tucker. Chichester, UK: John Wiley & Sons, Ltd.

Uthaman, Sudha, R.P. George, Vinita Vishwakarma, D. Ramachandran, B. Anandkumar, and U. Kamachi Mudali. 2019. "Enhanced biodeterioration resistance of nanophase modified fly ash concrete specimens: Accelerated studies in acid producing microbial cultures." *Environmental Progress & Sustainable Energy* 38 (2). John Wiley and Sons Inc.: 457–66.

Uthaman, Sudha, R. P. George, Vinita Vishwakarma, D. Ramachandran, C. Thinaharan, K. Viswanathan, and U. Kamachi Mudali. 2018. "Surface modification of fly ash concrete through nanophase incorporation for enhanced chemical deterioration resistance." *Journal of Bio- and Tribo-Corrosion* 4 (2). Springer International Publishing: 1–10.

Vishwakarma, Vinita. 2020. "Impact of environmental biofilms: Industrial components and its remediation." *Journal of Basic Microbiology* 60 (3). Wiley-VCH Verlag: 198–206.

Wang, Hua, Lu Kwang Ju, Homero Castaneda, Gang Cheng, and Bi Min Zhang Newby. 2014. "Corrosion of carbon steel C1010 in the presence of iron oxidizing bacteria acidithiobacillus ferrooxidans." *Corrosion Science* 89 (C). Elsevier Ltd: 250–57.

Wei, Shiping, Zhenglong Jiang, Hao Liu, Dongsheng Zhou, and Mauricio Sanchez-Silva. 2013. "Microbiologically induced deterioration of concrete - A review." *Brazilian Journal of Microbiology*. Sociedade Brasileira de Microbiologia 44(4): 1001–1007.

Xu, Congmin, Yaoheng Zhang, Guangxu Cheng, and Wensheng Zhu. 2007. "Localized corrosion behavior of 316L stainless steel in the presence of sulfate-reducing and iron-oxidizing bacteria." *Materials Science and Engineering A* 443 (1–2). Elsevier: 235–41.

Zhou, Li, Hai Ming Wong, and Quan Li Li. 2020. "Anti-biofouling coatings on the tooth surfacase and hydroxyapatite." *International Journal of Nanomedicine* 2020(15): 8963–8982. Dove Medical Press Ltd.

2

Biofouling in Industrial Water Systems, Membrane Biofouling: Assessment and Reduction Strategies

Sanghyun Jeong, Nirenkumar Pathak, Gayathri Naidu, and Saravanamuthu Vigneswaran

CONTENTS

2.1 Introduction

Biofouling is an inevitable phenomenon demonstrated by the attachment and buildup of microorganisms and the development of a biofilm (She et al. 2016) on the inner surfaces of pipelines and on membranes used in water treatment and desalination (Wingender, Neu, and Flemming 1999). The biofouling phenomena and remedies are the same for pipelines and reverse osmosis membranes used in seawater desalination and wastewater treatment for reuse. Two other chapters in this book discuss on biofouling in pipelines and their characterization methods. Thus, this chapter discusses mainly on biofouling in reverse osmosis membranes. This chapter aims to elucidate the biofouling mechanisms and their adverse effects, biofouling detection, and remediation methods. It also highlights the key issues related to the use of pretreatment schemes for biofouling mitigation.

The researchers reported that within a few years of the service, potable water distribution pipelines were found to be accumulated with a fine film of microbes that also pose major public health issues (van der Kooij, Visser, and Hijnen 1982). The microorganisms can regrow on the water-carrying pipe network in the presence of certain most significant constituents in water such as biodegradable organic matter (BOM), ammonia, iron, manganese, nitrite, soluble hydrogen, and reduced sulfur compounds. Moreover, it makes water biologically unstable due to the biofilm development caused by biodegradable organic matter (BOM), ammonia, iron, manganese, nitrite, dissolved hydrogen, and sulfur in reduced form. Water quality often adversely is affected by chlorine dosing that produces undesirable disinfection by-products (DBPs), which are unsafe and carcinogenic in nature.

In the seawater desalination process, RO membrane fouling is a major hurdle that reduces permeate flux and increases operating cost OMBRs (Sun et al. 2018). RO membrane commonly encounters colloidal fouling, organic fouling, inorganic scaling, and biofouling (Matin et al. 2011). The deposition of colloidal particles on membranes is called colloidal fouling and deposition and adsorption of macromolecular organic compounds on membranes is termed as organic fouling. The precipitation of dissolved inorganic compounds on the membrane surface is called inorganic scaling, while biofouling is the adhesion and accumulation of microorganisms on the membrane surface (She et al. 2016; Vrouwenvelder et al. 2009). The complex biofouling phenomena occur at the inner surface of pipes and on the RO membrane surface, which is accompanied by the agglomeration of soluble microbial products (SMP) and extracellular polymeric substance (EPS), thereby forming biofilm on the respective surface (Wingender, Neu, and Flemming 1999; Chun et al. 2017).

In response, the feed water pretreatment (such as microfiltration, ultrafiltration, chlorine dosing, and biocides addition) is highly effective in the elimination of 99.99% of microbes. Though only a few colonies of bacteria enter the system, they stick to the surfaces and then reproduce and grow on the surfaces in contact with even an oligotrophic environment (Chun et al. 2017). Therefore, biofouling is significant and inevitable in the RO membrane process even after periodic cleaning cycles and adapting other antifouling pretreatment strategies such as ozone and chlorine dosing or biocides application (Flemming et al. 1997). Furthermore, the polyamide RO membrane is susceptible to oxidation by free chlorine species (HOCl and OCl$^-$). However, the growth of resilient strains of microbes can be adversely affected by the constant use of disinfectants (Kang et al. 2007; Shannon et al. 2008). Biofouling leads to a drop in permeate velocity, selectivity, and membrane service life. Furthermore, it increases cleaning frequency and operational cost of chemicals and electricity (Luo et al. 2018; Vrouwenvelder, van Loosdrecht, and Kruithof 2011; Flemming et al. 1997).

In RO desalination, plant fouling is a major concern that requires frequent cleaning, shortens the membrane life, reduces permeate flow, and increases up to 50% total operating cost (Ridgway 2003; Bell, Holloway, and Cath 2016). Different cleaning tactics include the use of antiscalants and acids and importantly all such chemicals act very differently, and even certain commercially available chemicals cause biofouling (Vrouwenvelder et al. 2000).

2.2 Biofouling Mechanisms

Biofouling is a thin and compact gel-like biofilm layer formed on the inner surface of pipes or on the RO membrane surface. This biofilm formation involves three subsequent phases as shown in Figure 2.1.

i. Movement of microbes to the membrane surface,
ii. Adhesion to the surface, and
iii. Formation of nuclei and layer-by-layer addition of microbes on the surface (Al-Juboori and Yusaf 2012).

Biofilm formation in the membrane process is separate from the simple deposition of particles on the membrane, which are readily removed by physical washing (Miura, Watanabe, and Okabe 2007).

The first stage which occurs in minutes to hours is a reversible process. The initial attachment of organic matter, colloids, nutrients, and bacterial cells onto a membrane surface occurs at the time of contact of feed water and RO membrane (Subramani and Hoek 2010). During the second stage, cell attachment and micro-colony formation begin on the membrane surface where attached bacteria consolidate the bonding by secreting soluble microbial products that form complex with organic constituents. Consequently, biofilm adhesion becomes irreversible, and the organism becomes attached to the surface in a stack. This

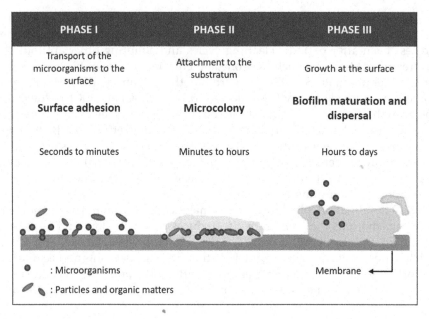

FIGURE 2.1
Schematic of biofilm formation.

phenomenon happens so rapidly when feed water meets the membrane skin layer. This biofilm is extremely hard to disengage from the membrane surface, which demands rigorous feed water pretreatment prior to its contact with the membrane (Adham et al. 1991; Habimana, Semião, and Casey 2014).

The adhesion of microbes onto the membrane surface is caused by complex physicochemical and biological processes. Feed characteristics, temperature, membrane properties, and module geometry play a role in this (Miura, Watanabe, and Okabe 2007; Subramani and Hoek 2010). The feed characteristics include pH, TDS, conductivity, and dissolved organic carbon. The properties of the membrane and pipes such as surface roughness, charge, and hydrophobicity are also significant parameters. The water flux is also an important factor in cell attachment to the membrane surface; however, the microbial community can attach via Brownian deposition in the absence of permeation (Subramani and Hoek 2010; Schneider et al. 2005).

2.3 Biofouling Control and Prevention Strategies

Biofouling poses major challenges, and its monitoring and control incur a huge financial burden on the RO water plants (Flemming 2002). Biofouling can be minimized by fouling control and prevention techniques. In the biofouling control strategy, chemical cleaning is performed to reestablish permeate flow. This can be achieved by increasing cleaning cycles. However, the frequency of chemical cleaning must be minimized as it increases the operating cost of chemicals and decreases the membrane life (Onoda 2016).

On the other hand, biofouling prevention is a more suitable option as compared to fouling control. Two common strategies for biofouling prevention are:

1. Membrane modification
2. Feed water pretreatment

Membrane modification technically controls the adhesion of microbes or inactivates the bacteria that get attached to the surface. This technique is still in the experimental stage and not explored yet for fouling prevention. On the other hand, the second alternative of feed pretreatment removes microbes, organics, and nutrients from the feed water. To achieve this objective, various pretreatment techniques are employed. They are MF/UF membrane filtration, ultraviolet light treatment, hypochlorite disinfection, or biocide use prior to RO filtration. The prevention techniques are beneficial as they incur less cost and energy consumption due to fewer amounts of chemicals used and lessen the adverse impact on the ecosystem. Furthermore, seawater desalination by the RO process becomes smooth and economical as feed pretreatment produces consistent water quality with a negligible amount of potential foulants to RO inlet. Raw water characteristics vary with geographical location and influent water quality analysis is essential to quantify biofouling potential. As a result, it is important to visualize and characterize the spatial distribution and transport of key membrane foulants. These fouling characteristics provide insights into the design of the pretreatment process and fouling mitigation strategies (Adham et al. 1991; Luo et al. 2018; Chun et al. 2017).

The design and operation of seawater reverse osmosis desalination (SWRO) processes relies heavily on the feed water characteristics. Thus, SWRO plant performance and selectivity are governed by the pretreated RO influent. Hence, the pretreatment process is a very significant section of the desalination plant. In this pretreatment stage, all particulate matters, colloidal particles, organics, nutrients, scalants, and bacterial impurities are removed from saline water in order to prevent them from reaching the sensitive RO membrane skin layer and prevent membrane fouling. The peculiarities and number of contaminants in source saline water (influent) directly relate to the pretreatment processes, produced water yield, and selectivity.

Preston (2005) reported that dissolved organic carbon (DOC) in seawater varies from 1 to 3 mg/L. The seawater analysis data can be obtained by performing laboratory analysis for certain target compounds (Lu and Wang 2019). Also, instrument analysis with liquid chromatography-organic carbon detection (LC-OCD) is based on the size-exclusion chromatography. It is utilized to characterize the water-soluble organic carbon (OC) and it provides

MW fractions of organic matters in seawater in terms of biopolymers including polysaccharides and proteins (>20,000 Da), humic substances (highly UV-absorbable, hydrophobic, 800–1,000 Da), building blocks (Breakdown products of humic substances, 350–600 Da), low molecular weight (LMW) acids (aliphatic and low molecular weight organic acids, biogenic organic matter, 350 Da), and low molecular weight (LMW) neutrals (alcohols, aldehydes, ketones, amino acids, biogenic organic matter, 350 Da) (Huber et al. 2011).

2.4 Pretreatment and Membrane-Based Systems

Leparc et al. (2007) reported that for seawater pretreatment, conventional treatments such as dual media filtration (DMF) and cartridge filters and advanced pretreatments such as MF/UF filtration are employed (Leparc et al. 2007). The advanced pretreatment is getting more acceptance due to the complete removal of seawater contaminants as compared to DMF. Furthermore, few chemicals are consumed and MF/UF-based membrane filtration techniques are robust. MF/UF-based membrane filtration is gaining more popularity and acceptance as a pretreatment to RO. MF/UF provides consistent quality of safe and secure water to RO due to its absolute rejection for suspended solids irrespective of source seawater quality (Baig and Al Kutbi 1998; Pearce 2007). MF filters are available in pore sizes of 0.1–0.2 μm. UF membranes possess fine pores as compared to MF. UF is available in the range of 0.01–0.02 μm and sometimes even smaller to 0.005 μm. UF membranes are capable to reject all particulate matters and the majority of dissolved substances, including bacteria and viruses. The UF has better rejection than MF due to its low molecular weight cut-off (MWCO). The MF/UF rejection is subjected to the feed water characteristics and molar mass of the species (Baig and Al Kutbi 1998).

2.4.1 Biofilters

Biofilters are extensively used in air, water, and domestic sewage treatment characterized by biomass attached to its septum. A variety of biofilters are being used in water/wastewater treatment applications such as trickling filters, granulated activated carbon (GAC), sand filters, and horizontal rock filters to name a few. GAC-based biofilters were found useful typically in potable water purification due to the re-growth of bacterial colonies in a water distribution network pipe.

GAC-based biological treatments are more effective in oxidizing organic matters responsible for bacterial growth in a drinking water pipe. GAC biofilters are effective in water purification specifically after disinfection by ozone. Organic substances present in water impair water quality by imparting odor and taste that are esthetically undesirable. Also, organic micropollutants and organic precursors responsible for disinfection by-products (DBP) formation are undesirable.

2.5 Biofouling Reduction Strategies

As stated in the previous section, biofouling strategies are biofouling control and biofouling prevention. Chemical cleaning of the membrane is the major biofouling control technique. The biofouling prevention technique potentially includes feed water pretreatment, RO membrane surface modification, and water disinfection.

2.5.1 Direct Methods

In 'direct methods', biofouling is controlled from the time when membrane modules are manufactured. The immediate method is to control biofouling in situ by applying cleaning chemicals to the membrane directly. RO membrane life can be enhanced, and operational costs can be minimized by following appropriate cleaning protocols.

2.5.2 Modification

As a biofouling prevention strategy, antifouling agents are incorporated into the membrane during the preparation stage. Thus, membrane modification improves its physicochemical properties, thereby reducing biofouling. The membrane properties to be improved for fouling prevention includes functional group, smoothness, charge, and hydrophobicity (Louie et al. 2006; Chae et al. 2009; Malaisamy et al. 2010; Miller et al. 2012).

By enhancing membrane surface properties, biomass adhesion and attachment can be minimized and the bacterial species becomes inactive. Rough membrane surfaces are prone to increase biomass attachment than smooth surfaces (Louie et al. 2006). Microbes in the aqueous phase possess negative charge and so negatively charged membranes repel each other and thus negative charge on a membrane surface helps in preventing biofouling (Hori and Matsumoto 2010). However, negatively charged foulants are attracted to the membrane surface.

The hydrophilic and hydrophobic nature of the membrane is also linked to the membrane fouling. Kwon et al. (2005) reported that the more the hydrophilicity of the membrane, the lesser the degree of biofouling. However, a negatively charged organic foulant can easily attach to the membrane. Membrane modification with a chemical surfactant is also a promising option. The bio-surfactants are alternative to the chemical surfactants, which are derived from renewable materials. Bio-surfactants possess less inter-facial and surface tensions in both water phase and organic solutions. The use of bio-surfactants is beneficial because they are biodegradable in nature, environmentally friendly, compatible under very high temperature, pH, and saline conditions, less toxic, and have high form formation and selectivity (Desai and Banat 1997; Wilbert, Pellegrino, and Zydney 1998).

One of the limitations of using bio-surfactants is the high cost. Furthermore, in the beginning, microbes may not attach to the membrane surface in the presence of bio-surfactants but in the longer run, microbes develop adapt-ability to grow under hostile environmental conditions (Hori and Matsumoto 2010). To achieve the best biofouling prevention using bio-surfactants, feed water must be free from any living biomass.

2.5.3 Cleaning

The strategies of membrane cleaning are plentiful and normally remain pro-prietary. The literature report suggests that permeate flux can be restored by adopting a proper cleaning protocol (Madaeni and Mansourpanah 2004). Membrane cleaning is recommended when the transmembrane pressure increases by 10% or the flux value drops by 10%. The major parameters that affect the chemical cleaning performance are the type of chemical selected for cleaning and its concentration, duration of cleaning, pH, and temperature (Al-Amoudi and Farooque 2005). Fane (1997) reported that about 5%–20% of the operating cost component is consumed toward membrane cleaning in an RO desalination plant.

Biofilms are attached to the membrane surface and to disengage the bio-film, two main techniques are used.

1. Application of air scouring or high crossflow velocity (high shearing velocity)
2. Detachment of the biofilm employing proper chemicals (Fleming 2002)

As a physical cleaning strategy, backwashing and/or relaxation are the commonly used techniques. These methods are performed on a regular basis, but their effectiveness will be reduced by filtration time. When irre-versible fouling is observed on the surface ,different intensities of chemical cleanings can be injected on a regular basis (weekly to yearly). In chemi-cal cleaning techniques, chemicals inhibit the microbial bonding with the membrane surface and loosen them. Sodium hydroxide (NaOH) (Kim et al. 2011) and sodium hypochlorite (NaOCl) (Subramani and Hoek 2010) are gen-erally used cleaning chemicals. NaOH demonstrated excellent performance removing 95% biofouling from membrane when cleaning was performed for 20 minutes. NaOCl (0.3%) is usually utilized as a major chemical agent in the microfiltration membrane processes to remove the organic foulants, while citric acid is normally used for inorganic foulants (Le-Clech 2010). However, chemical cleaning could not absolutely remove the attached biofilm, and fast regrowth of biomass was observed (Kim et al. 2009; Vrouwenvelder et al. 2003; Bereschenko et al. 2011). Moreover, the thin polyamide skin layer gets damaged with cleaning agents. Also, the production of less amount of pure water and disposal of chemical waste are other issues that need to be tackled (Khan et al. 2010; Kang et al. 2007).

2.5.4 Indirect Methods

MF/UF-based membrane filtration and biofiltration are used as indirect pretreatment techniques to RO feed. Those pretreatment methods ensure that organics, nutrients, particles, and microbes are physically separated from raw water. Sometimes sodium bisulfite-based biocides and NaOCl as disinfectants are used to destroy microbes.

2.5.5 Pretreatment

Pretreatment is crucial in SWRO because it reduces the biofouling occurrence and thereby improves RO desalination plant efficiency. To achieve this, suspended solids, organics, nutrients, minerals, bacteria, and trace organics are removed from the source water (Kumar and Sivanesan 2006). Thus, RO feed water quality is improved, and it becomes free from biomass. This minimizes bacterial tendency to attach to the RO membrane. RO feed water treatment can be done either by employing conventional physicochemical methods or by recently used membrane-based separation. Leparc et al. (2007) reported that dual media filtration (DMF) and cartridge filters could not completely reject source seawater contaminants. The advanced pretreatment is getting more acceptance and MF/UF membranes could achieve better removal of harmful impurities as compared to DMF. Furthermore, few chemicals are consumed and MF/UF-based membrane filtration techniques are robust. MF membranes are fabricated from polyvinylidene fluoride (PVDF) as they provide mechanical strength and resist chemical attack (Ding et al. 2013). The MF membrane rejects suspended and dissolved particles, but absolute elimination of bacteria is not obtained. The MF process operates around 2 bar pressure, which is usually lower than that of nanofiltration (NF) and UF (Sachit and Veenstra 2014).

2.5.5.1 Deep Bed Biofilter (DBF)

Deep bed filters behave like a biofilter when they are operated at low filtration rates and allow the formation of a biofilm on their surface. When GAC is used as a filter medium, the process begins with sorption of organic molecules onto the filter media followed by enzymatic hydrolysis (enzymes secreted from bacteria are attached to the biofilm) of bigger molecules to the tiny fractions. Those small fractions then transport to the biofilm, which further metabolizes the biodegradable organics and consumes as a substrate from the feed water (Hu et al. 2005; Larsen and Harremoës 1994).

Naidu et al. (2013) examined the granular activated carbon (GAC) biofilter performance in biomass adhesion on its surface when treating seawater. Biomass activity was measured in terms of ATP and the bacterial population was expressed in CFU. The biomass accumulation and DOC removal were correlated. The authors reported that within 20 days of experimental operation, a high amount of bacterial mass (1.0×108 CFU/g media) was deposited on the top surface of the biofilter. They observed that with decreasing thickness of filter media, biomass accumulation was reduced. Moreover, in the early phase of the study, the microbial concentration was 0.9 ± 0.5 µg ATP/g media within 0–5 days period and after reaching the steady state within 15–20 days, it increased to 51.0 ± 11.8 µg ATP/g media and the filter produced good quality of treated water (the DOC concentrations were 0.51 ± 0.12 mg/L). It was reported that compared to sand filter and anthracite, the GAC performed better and this can be attributed to the high porosity of GAC capable of providing more surface area and accumulation of higher biomass (Wang, Summers, and Miltner 1995).

In another study, Jeong et al. (2013) evaluated the performance of GAC and anthracite biofilter in seawater desalination. Terminal restriction fragment length polymorphism (T-RFLP), principal component analysis (PCA), and 16S rRNA gene sequencing techniques were used for bacterial consortia analysis. The authors deduced that the GAC biofilter captured diverse heterotrophs and outperformed the anthracite filter during 75 days of operation. High AOC removal was linked to the abundance of the microbial community on GAC. When the process was in the early phase of the operation, effluent AOC concentration was high (18.0 ± 1.4 µg-C glucose/L). After attaining a steady state within a 15–20 days period, AOC in the effluent was low (0.6 ± 0.2 µg-C glucose/L). The high AOC in the early phase was linked to the higher molar mass of the organics, which assimilated to the low molar mass fractions. Once the steady state is attained, the specific microbial consortia proliferated over time and metabolized the low molecular weight organic reducing AOC in the permeate (Naidu et al. 2013). On the other hand, the anthracite biofilter was selective for sulfur-oxidizing and reducing bacteria. This can be attributed to the sulfur availability in the anthracite as a contaminant.

Though biofilter is an attractive pretreatment alternative, it has some limitations. When biofilter is put in service, it requires some time for acclimation. In this acclimation phase, certain microbes and nutrients pass through the biofilter and form a colony on the membrane surface. Similar phenomena were observed during the backwashing cycle (Sadr Ghayeni et al. 1998; Chua, Hawlader, and Malek 2003). Furthermore, for a biofilter to operate efficiently, some vital factors need to be considered such as filter media, feed flowrate, and cleaning cycles.

2.5.5.2 *Membrane Filtration*

Membrane filtration is another effective pretreatment since commercial membrane produces high water throughput and is cheaper. Furthermore, the small footprint of the membrane plant and few inventory requirements are the benefits of using membrane pretreatment.

The microbes can be rejected by employing membrane filtration and researchers suggested that this bacterial removal mechanism is the combination of two processes: (i) the effect of physio-chemical interactions between the membrane and microorganisms and (ii) the sieving effect (Košutić and Kunst 2002; Van der Bruggen et al. 1999). A membrane typically rejects the larger diameter of bacteria from passing through it. Also, the negatively charged membrane and microbes repel each other.

The membrane pretreatment potentially rejects nutrients from the raw water and thus prevents biofilm growth on the downstream RO system. As such, microbes do not receive enough nutrients from the feed. This malnourishment condition adversely affects reproduction and proper growth of microbes resulting in a thinner and unevenly distributed biofilm with less biofouling potential (Al-Juboori and Yusaf 2012; Flemming et al. 1997).

2.5.6 Biochemical Methods

Biochemical techniques such as bacteriophage, signaling molecules, and enzymes are employed to remove a rigidly attached biofilm on the surface (Flemming 2011). Bacteriophages are viruses that kill bacteria (Fu et al. 2010). The recently invented quorum sensing method uses extremely specific signaling biomolecules, which deliberately destroy cell–cell communication in microbes of the biofouling layer (Davies and Marques 2009). Although quorum sensing is a very promising biofouling removal method, it suffers from certain downsides such as high cost associated to process such biochemical molecules on a commercial basis and lack of consistency and efficacy in removing attached biomass using such methods (Richards and Cloete 2010; Flemming 2011).

2.5.7 Water Disinfection Method

Introducing disinfectants prior to the RO membrane is proved to be a very efficient pretreatment technique that simply prevents bacterial bonding onto the membrane surface (Hori and Matsumoto 2010). Disinfection methods include both chemical and thermal means as a conventional method and by application of such pretreatment, microbes are destroyed at the source and thus prevent them from reaching out to the membrane surface. Non-conventional treatments such as ultraviolet (UV) light treatment, mechanical treatment, and ultrasound treatment are also being employed.

Chemical pretreatment employs chemicals such as hypochlorite (Cl_2 based) and ozone molecules, which are generated in situ using an ozonizer. These disinfectants have a very wide potential to destroy microbes and are cheaper though the formation of disinfection by-products (DBPs) is a limitation with this process.

In this regard, solar energy (Davies et al. 2009) is considered another attractive alternative to the chemical pretreatment. It is one of the most promising economical pretreatments that suffers from limitations of low efficiency due to varying topographical and weather conditions. Specifically, during night-time, solar energy is not available, so solar system efficiency falls to zero and storage is not much viable alternative techno-economically (Davies et al. 2009).

UV light source is also used as a disinfectant technique (Schwartz, Hoffmann, and Obst 2003). The high cost of a UV lamp, energy consumption, DBP formation are some of the issues associated with the application of UV light. Furthermore, water characteristics such as turbidity and color adversely affect UV light performance due to absorption and Tyndall effects in the aqueous phase (Harris et al. 1987; Parker and Darby 1995).

Ultrasound has emerged as an attractive ecofriendly pretreatment option. Ultrasound has the ability to destroy the microbes and detach the biomass from the membrane surface (Gogate and Kabadi 2009; Joyce et al. 2003).

2.6 Detection of Fouling

The most significant factor that contributes to the decline in RO performance in the desalination process is the fouling of the membranes caused by adsorption and accumulation of particulate and organic foulants into the pores and onto the membrane surface (Inaba et al. 2017). Fouling impedes the effectiveness and throughput of RO by declining water flux, membrane selectivity, and permeate quality (Luo et al. 2018; Ng and Elimelech 2004). The root cause and major fouling potential to the membrane process are feed water characteristics (Vrouwenvelder et al. 2003) though operating parameters such as operating flux and recovery are also contributing factors (Chen et al. 2004).

2.6.1 Organic Matter (OM) in Seawater

Organic contaminants of the feed water lead to the organic fouling, which combines with other foulings such as colloidal and biofouling to contribute to the overall fouling in the system. Biofouling is seen as a living form of organic fouling and organic matters are believed to be a non-living form of biofouling resulting from bacterial metabolisms and its cellular fractions (Amy 2008) (Table 2.1).

TABLE 2.1

Various OM Measurements and Characterization of Feed Water Samples

Measurement Category	Protocol
Molecular weight (MW) distribution by size exclusion chromatography with online DOC detection (SEC–DOC)	OM in terms of chromatographic peaks corresponding to high molecular weight (MW) polysaccharides (PS), medium MW humic substances (HS) consisting of humic and fulvic acids, and low MW acids (LMA); this technique is conceptually equivalent to LC–OCD, liquid chromatography with organic carbon detection
Hydrophobic/transphilic/hydrophilic (HPO/TPI/HPI) DOC distribution	XAD-8/XAD-4 resin adsorption chromatography, revealing a polarity distribution of OM
3-Dimensional spectra fluorescence excitation–emission matrix (3D-FEEM)	Distinguishing between humic-like and protein-like OM as well as providing a fluorescence index (FI) that is related to the OM source

Source: Modified from Amy (2008).

2.6.2 Parameters Characterizing Biomass

Total direct cell count (TDC), adenosine 5'-triphosphate (ATP), and heterotrophic plate count (HPC) are significant parameters to measure biomass (Vrouwenvelder and van der Kooij 2001). The epifluorescence microscopy is used for TDC measurements with different dyes such as SYTO, acridine orange, and 4-6-diamidino-2-phenylindole (DAPI). The downside is that those dyes unfortunately stain the whole microbial communities available in the specimen. In other words, it stains both living and non-living cells. On the contrary, ATP is a more reliable, rapid, and easy method of biomass measurement that only considers and senses living cells. The active cells can be determined by light production through the enzymatic process by means of luciferin and firefly luciferase. The ATP amount and light produced had a linear relationship that determines ATP concentration. In the HPC analysis, samples are kept at 20°C or 28°C for an incubation period of 5–7 days on R2A plates to acquire heterotrophic bacterial cell counts in the mixed liquor. However, all the above-mentioned methods have limitations in assessing biomass when cells are in the cluster form. In this regard, still ATP is a more reliable, precise, and distinct method. Certainly, biomass detection in feed water and on the membrane employs a combination of ATP and TDC (Vrouwenvelder et al. 2008).

The choice of the pretreatment process largely depends on the quantity and the diversity of bacterial consortia (Schneider et al. 2005). A wide and diverse bacterial community is reported in saline water. It has been reported that certain groups of bacteria or dominant species are responsible for high molecular weight organics concentration or SMP (polysaccharides/protein) secretion (Frias-Lopez et al. 2002; Cottrell and Kirchman 2000).

2.6.3 Fouling Potential of Water

2.6.3.1 Particulate Fouling Potential

To measure and detect particulate fouling in feed water and on the membrane surface, a suitable fouling detection technique is essential (Boerlage et al. 2003). Particulate fouling details are useful during the design of the entire plant specifically for pretreatment. Also, this is significant to monitor plant performance and efficiency.

The particulate matters fouling can be measured and indicated by the silt density index (SDI) and modified fouling index (MFI). In the beginning, MFI with 0.45 μm filter media was employed for particulate matters measurement in feed water. Later, Moueddeb, Jaouen, and Schlumpf (1996) pointed out shortcomings of this method and then Boerlage et al. (2003) established a novel UF-MFI. UF-MFI is a promising technique in fouling characterization for a certain source of feed water and records any variation in RO feed water characteristics (Boerlage et al. 2003).

2.6.3.2 Extracellular Polymeric Substances

The amount of SMP and EPS significantly affects biofouling. Both are heterogeneous in nature and consist of a variety of organics mainly polysaccharides, proteins, humic acid, glycolipids, and deoxyribonucleic acid (DNA) (Wang et al. 2014). Thus, membrane biofouling is a dynamic and slow process revealed by an addition of self-originated microbial cells to the membrane surface by glue-like, autogenic soluble microbial products (Inaba et al. 2017). EPS originated from biomass are the main component that contributes to biofouling and causes membrane permeability decline with time (Wang et al. 2016).

Berman (2010) deduced that transparent exopolymer particles (TEP) are a major contributor to the biofouling on a RO membrane. The TEP role is very vital in bacterial growth and this gluelike TEP layer helps in biofilm formation over the RO membrane. In another report, it has been reported that about 68% of the total microbial community were attached to the TEP component and this TEP measurement and monitoring is very useful to understand RO fouling phenomena (Villacorte et al. 2009).

2.6.3.3 Biofouling Potential

The omnipresence of bacteria and the amount of nutrients actually decide the growth of biomass and thus biofilm formation on the membrane surface. The biofilm formation growth depends on various operating parameters such as shear velocity, nutrient type (N, P, K), concentration, and robustness of the biofilm attached to the surface (Flemming 1997). The cleaning chemicals usage such as biocides and disinfectant quality directly depends on the biofouling detection (Vrouwenvelder et al. 2000; Vrouwenvelder, van Loosdrecht, and Kruithof 2011; Vrouwenvelder et al. 2007).

The biodegradable organic matter (BOM) is measured as assimilable Organic Carbon (AOC) and biodegradable dissolved organic carbon (BDOC) and the BOM is a limiting constituent for biomass growth (LeChevallier, Schulz, and Lee 1991). The AOC component of BOM is linked to the microbial population in feed water (Hambsch and Werner 1996; Weinrich, Schneider, and LeChevallier 2011) and it represents low molecular weight compounds such as acetic acids and amino acids.

2.6.3.4 Assimilable Organic Carbon (AOC)

The assimilable organic carbon (AOC) is typically 0.1%–9.0% of the TOC, which can easily be taken up by microbes for their metabolism and growth. This biomass growth is measured by the colony count in AOC analysis. For this, a standard concentration plot is prepared to show microbial growth yield and assimilable organic concentration. The growth monitored during the incubation is converted to the AOC from the standard curve. van der Kooij in 1992 reported that when AOC was <10 µg/L, the heterotrophic bacterial growth was limited. Based on van der Kooij's (1992) concept, many analytical techniques were developed to estimate AOC. Table 2.2 presents the representative AOC analysis techniques. Furthermore, some of these methods deviate from adapting natural bacterial consortia (Hammes and Egli 2005; Kaplan, Bott, and Reasoner 1993), rather than using pure cultures. Moreover, a majority of AOC available uses various growth measuring methods such as plating, ATP, turbidity, flow cytometry, and luminescence. The current research efforts are aiming at user-friendly and rapid AOC detection methods development.

Jeong et al. (2013) introduced a novel AOC measurement technique called the *Vibrio fischeri* method. In this method, frozen *V. fischeri* stock is allowed for incubation for a short time in seawater. Glucose is used as a carbon supply. The luminescence meter measures natural bioluminescence after the incubation period at 25°C wherein marine agar plate is employed for strain preparation. The distinctive feature of this method is the very less (30 minutes) incubation time due to the directly used marine agar plate for strain preparation. Moreover, *V. fischeri* strain outperformed the previously used *V. harvey* strain for AOC detection due to its good correlation with cell number and luminescence. Furthermore, as *V. fischeri* strain was derived from seawater, it can therefore be well adopted to high TDS concentration of source saline water. The luminescence detection approach is also favorable to flow cytometry as luminescence detection is much easier. Flow cytometry suffers from the limitation of detecting too small cell counts (<10^2 cells/mL).

2.6.3.5 Biodegradable Organic Carbon (BDOC)

The portion of available dissolved organic carbon (DOC) of the feed, which is easily metabolized by heterotrophic bacteria, is biodegradable dissolved organic carbon (BDOC) (Servais, Billen, and Hascoët 1987). The BDOC is

TABLE 2.2

Representative AOC Methods Available in the Literature

Methods	Target (volume, mL)	Culture	Incubation Time (days/h)	Cell Counts	Substrate	Detection Limits	Ref.
Van der	Drinking	*Pseudomonas fluorescens*	7–9	Nutrient agar	Acetate	10 µg/L	van der Kooij et al. (1982)
Eawag	Tap water	Precultured natural	2–3	Flow cytometer	Acetate	10 µg/L	Hammes and Egli (2005)
V. harvey	Saline water	*V. harvey*	<1h	Luminescence	Acetate	<10 µg/L	Weinrich et al. (2011)
V. fischeri	Seawater	*V. fischeri*	<1h	Luminescence	Glucose	0.1 µg-C	Jeong et al. (2013a)

Source: Modified from Jeong et al. (2019).

obtained by subtracting the initial DOC from the final DOC detected after 28 days of incubation. For incubation, the inoculum chosen is simply environmental microbes. The incubation happens under suspended and attached growth conditions. For the attached growth mechanism, sand or porous bed supports are provided. During suspended incubation, 28 or 5–7 days are recommended time for bacteria attached to the sand support. The BDOC detection is a biodegradability indication parameter generally used in water treatment. Nevertheless, van der Kooij (1992) deduced that due to lack of proper correlation between bacteria and BDOC concentration, BDOC is not reliable in accurately predicting bacterial regrowth, and due to its extremely low detection limit (0.1 mg/L), it also measures AOC concentration (van der Kooij 1992).

The Biomass Production Potential (BPP) test is performed when the biodegradable chemical of a water sample cannot be consumed by the AOC test. BPP can detect maximum ATP of the microbial community of the water under 25°C incubation. BPP is measured in terms of ATPmax/mg product or liter of water (Vrouwenvelder et al. 2000). Another such parameter, the biofilm formation rate (BFR), is expressed as pgATPcm^2/day, which is nothing but biotic bacteria (ATP) deposition on the glass ring surface measured by an online biofilm monitor (van der Kooij et al. 1995).

References

Adham, Samer S., Vernon L. Snoeyink, Mark M. Clark, and Jean-Luc Bersillon. 1991. "Predicting and verifying organics removal by PAC in an ultrafiltration system." *American Water Works Association* 83 (12). John Wiley & Sons, Ltd: 81–91.

Al-Amoudi, Ahmed S., and A. Mohammed Farooque. 2005. "Performance restoration and autopsy of NF membranes used in seawater pretreatment." *Desalination* 178 (1–3 SPEC. ISS.). Elsevier: 261–71.

Al-Juboori, Raed A., and Talal Yusaf. 2012. "Biofouling in RO system: Mechanisms, monitoring and controlling." *Desalination*, 302:1–23. Elsevier.

Amy, Gary. 2008. "Fundamental understanding of organic matter fouling of membranes." *Desalination* 231 (1–3). Elsevier: 44–51.

Baig, M. B., and A. A. Al Kutbi. 1998. "Design features of a 20 Migd SWRO desalination plant, Al Jubail, Saudi Arabia." *Desalination* 118 (1–3). Elsevier Sci B.V.: 5–12.

Bell, Elizabeth A., Ryan W. Holloway, and Tzahi Y. Cath. 2016. "Evaluation of forward osmosis membrane performance and fouling during long-term osmotic membrane bioreactor study." *Journal of Membrane Science* 517 (November). Elsevier B.V.: 1–13.

Bereschenko, L. A., H. Prummel, G. J.W. Euverink, A. J. M. Stams, and M. C. M. van Loosdrecht. 2011. "Effect of conventional chemical treatment on the microbial population in a biofouling layer of reverse osmosis systems." *Water Research* 45 (2). Elsevier Ltd: 405–16.

Boerlage, Siobhan F. E., Maria D. Kennedy, Meseret Petros Aniye, Elhadi Abogrean, Zeyad S. Tarawneh, and Jan C. Schippers. 2003. "The MFI-UF as a water quality test and monitor." *Journal of Membrane Science* 211 (2). Elsevier: 271–89.

Chae, So Ryong, Shuyi Wang, Zachary D. Hendren, Mark R. Wiesner, Yoshimasa Watanabe, and Claudia K. Gunsch. 2009. "Effects of fullerene nanoparticles on escherichia coli K12 respiratory activity in aqueous suspension and potential use for membrane biofouling control." *Journal of Membrane Science* 329 (1–2). Elsevier: 68–74.

Chen, Kai Loon, Lianfa Song, Say Leong Ong, and Wun Jern Ng. 2004. "The development of membrane fouling in full-scale RO processes." *Journal of Membrane Science* 232 (1–2). Elsevier: 63–72.

Chua, K. T., M. N. A. Hawlader, and A. Malek. 2003. "Pretreatment of seawater: Results of pilot trials in singapore." *Desalination* 159 (3). Elsevier: 225–43.

Chun, Youngpil, Dennis Mulcahy, Linda Zou, and In Kim. 2017. "A short review of membrane fouling in forward osmosis processes." *Membranes* 7 (2). MDPI AG: 30.

Cottrell, Matthew T., and David L. Kirchman. 2000. "Natural assemblages of marine proteobacteria and members of the cytophaga-flavobacter cluster consuming low- and high-molecular-weight dissolved organic matter." *Applied and Environmental Microbiology* 66 (4). Appl Environ Microbiol: 1692–97.

Davies, C. M., D. J. Roser, A. J. Feitz, and N. J. Ashbolt. 2009. "Solar radiation disinfection of drinking water at temperate latitudes: Inactivation rates for an optimised reactor configuration." *Water Research* 43 (3). Elsevier Ltd: 643–52.

Davies, David G., and Cláudia N. H. Marques. 2009. "A fatty acid messenger is responsible for inducing dispersion in microbial biofilms." *Journal of Bacteriology* 191 (5). J Bacteriol: 1393–403.

Desai, J. D., and I. M. Banat. 1997. "Microbial production of surfactants and their commercial potential." *Microbiology and Molecular Biology Reviews : MMBR* 61 (1). American Society for Microbiology: 47–64.

Ding, Yi, Yu Tian, Zhipeng Li, Haoyu Wang, and Lin Chen. 2013. "Microfiltration (MF) membrane fouling potential evaluation of protein with different ion strengths and divalent cations based on extended DLVO theory." *Desalination* 331 (December). Elsevier: 62–8.

Flemming, H. C. 2002. "Biofouling in water systems - cases, causes and countermeasures." *Applied Microbiology and Biotechnology*, 59(6): 629–40.

Flemming, H. C., G. Schaule, T. Griebe, J. Schmitt, and A. Tamachkiarowa. 1997. "Biofouling - The achilles heel of membrane processes." *Desalination* 113 (2–3). Elsevier: 215–25.

Flemming, Hans Curt. 1997. "Reverse osmosis membrane biofouling." *Experimental Thermal and Fluid Science* 14 (4). Elsevier Inc.: 382–91.

Flemming, Hans-Curt. 2011. "Microbial biofouling: Unsolved problems, insufficient approaches, and possible solutions." 81–109. Biofilm Highlights, Springer: Berlin, Heidelberg.

Frias-Lopez, Jorge, Aubrey L. Zerkle, George T. Bonheyo, and Bruce W. Fouke. 2002. "Partitioning of bacterial communities between seawater and healthy, black band diseased, and dead coral surfaces." *Applied and Environmental Microbiology* 68 (5). Appl Environ Microbiol: 2214–28.

Fu, Weiling, Terri Forster, Oren Mayer, John J. Curtin, Susan M. Lehman, and Rodney M. Donlan. 2010. "Bacteriophage cocktail for the prevention of biofilm formation by pseudomonas aeruginosa on catheters in an in vitro model system." *Antimicrobial Agents and Chemotherapy* 54 (1). Antimicrob Agents Chemother: 397–404.

Gogate, Parag R., and Abhijeet M. Kabadi. 2009. "A review of applications of cavitation in biochemical engineering/biotechnology." *Biochemical Engineering Journal*, 44(1): 60–72. Elsevier.

Habimana, O., A. J. C. Semião, and E. Casey. 2014. "The role of cell-surface interactions in bacterial initial adhesion and consequent biofilm formation on nanofiltration/reverse osmosis membranes." *Journal of Membrane Science*, 454: 82–96. Elsevier.

Hambsch, B, and P. Werner. 1996. "The removal of regrowth enhancing organic matter by slow sand filtration." In *Advances in Slow Sand and Alternative Biological Filtration*, 21–7. Nigel Graham (Editor), Robin Collins (Editor). John Wiley & Sons. https://www.wiley.com/en-us/Advances+in+Slow+Sand+and+Alternative+Biological+Filtration-p-9780471967408.

Hammes, Frederik A., and Thomas Egli. 2005. "New method for assimilable organic carbon determination using flow-cytometric enumeration and a natural microbial consortium as inoculum." *Environmental Science and Technology* 39 (9). American Chemical Society : 3289–94.

Harris, George D., V. Dean Adams, Darwin L. Sorensen, and Michael S. Curtis. 1987. "Ultraviolet inactivation of selected bacteria and viruses with photoreactivation of the bacteria." *Water Research* 21 (6). Pergamon: 687–92.

Hori, Katsutoshi, and Shinya Matsumoto. 2010. "Bacterial adhesion: From mechanism to control." *Biochemical Engineering Journal*, 48(3): 424–434. Elsevier.

Hu, J. Y., L. F. Song, S. L. Ong, E. T. Phua, and W. J. Ng. 2005. "Biofiltration pretreatment for reverse osmosis (RO) membrane in a water reclamation system." *Chemosphere* 59 (1). Elsevier Ltd: 127–33.

Huber, Stefan A., Andreas Balz, Michael Abert, and Wouter Pronk. 2011. "Characterisation of aquatic humic and non-humic matter with size-exclusion chromatography - organic carbon detection - organic nitrogen detection (LC-OCD-OND)." *Water Research* 45 (2). Elsevier Ltd: 879–85.

Inaba, Tomohiro, Tomoyuki Hori, Hidenobu Aizawa, Atsushi Ogata, and Hiroshi Habe. 2017. "Architecture, component, and microbiome of biofilm involved in the fouling of membrane bioreactors." *NPJ Biofilms and Microbiomes* 3 (1). Nature Publishing Group: 5.

Jeong, Sanghyun, Gayathri Naidu, Saravanamuthu Vigneswaran, Chao Hoe Ma, and Scott A. Rice. 2013. "A rapid bioluminescence-based test of assimilable organic carbon for seawater." *Desalination* 317 (May). Elsevier: 160–65.

Joyce, E., T. J. Mason, S. S. Phull, and J. P. Lorimer. 2003. "The development and evaluation of electrolysis in conjunction with power ultrasound for the disinfection of bacterial suspensions." *Ultrasonics Sonochemistry*, 10. Elsevier: 231–34.

Kang, Guo Dong, Cong Jie Gao, Wei Dong Chen, Xing Ming Jie, Yi Ming Cao, and Quan Yuan. 2007. "Study on hypochlorite degradation of aromatic polyamide reverse osmosis membrane." *Journal of Membrane Science* 300 (1–2). Elsevier: 165–71.

Kaplan, L. A., T. L. Bott, and D. J. Reasoner. 1993. "Evaluation and simplification of the assimilable organic carbon nutrient bioassay for bacterial growth in drinking water." *Applied and Environmental Microbiology* 59 (5). American Society for Microbiology (ASM): 1532–39.

Khan, Mohiuddin Md Taimur, Philip S. Stewart, David J. Moll, William E. Mickols, Mark D. Burr, Sara E. Nelson, and Anne K. Camper. 2010. "Assessing biofouling on polyamide reverse osmosis (RO) membrane surfaces in a laboratory system." *Journal of Membrane Science* 349 (1–2). Elsevier: 429–37.

Kim, Dooil, Seunghoon Jung, Jinsik Sohn, Hyungsoo Kim, and Seockheon Lee. 2009. "Biocide application for controlling biofouling of SWRO membranes - an overview." *Desalination* 238 (1–3). Elsevier: 43–52.

Kim, Lan Hee, Am Jang, Hye Weon Yu, Sung Jo Kim, and In S. Kim. 2011. "Effect of chemical cleaning on membrane biofouling in seawater reverse osmosis processes." *Desalination and Water Treatment* 33 (1–3). Taylor and Francis Inc.: 289–94.

Košutić, K., and B. Kunst. 2002. "Removal of organics from aqueous solutions by commercial RO and NF membranes of characterized porosities." *Desalination* 142 (1). Elsevier: 47–56.

Kumar, K. Vasanth, and S. Sivanesan. 2006. "Pseudo second order kinetic models for safranin onto rice husk: Comparison of linear and non-linear regression analysis." *Process Biochemistry* 41 (5). Elsevier: 1198–202.

Kwon, Boksoon, Sangyoup Lee, Jaeweon Cho, Hyowon Ahn, Dongjoo Lee, and Heung Sup Shin. 2005. "Biodegradability, DBP formation, and membrane fouling potential of natural organic matter: Characterization and controllability." *Environmental Science and Technology* 39 (3). Environ Sci Technol: 732–39.

Larsen, Tove A., and Poul Harremoës. 1994. "Degradation mechanisms of colloidal organic matter in biofilm reactors." *Water Research* 28 (6). Pergamon: 1443–52.

LeChevallier, M. W., W. Schulz, and R. G. Lee. 1991. "Bacterial nutrients in drinking water." *Applied and Environmental Microbiology* 57 (3). American Society for Microbiology (ASM): 857–62.

Le-Clech, Pierre. 2010. "Membrane bioreactors and their uses in wastewater treatments." *Applied Microbiology and Biotechnology*, 88, 1253–1260. Springer Verlag.

Leparc, Jérôme, Sophie Rapenne, Claude Courties, Philippe Lebaron, Jean Philippe Croué, Valérie Jacquemet, and Greg Turner. 2007. "Water quality and performance evaluation at seawater reverse osmosis plants through the use of advanced analytical tools." *Desalination* 203 (1–3). Elsevier: 243–55.

Louie, Jennifer S., Ingo Pinnau, Isabelle Ciobanu, Kenneth P. Ishida, Alvin Ng, and Martin Reinhard. 2006. "Effects of polyether-polyamide block copolymer coating on performance and fouling of reverse osmosis membranes." *Journal of Membrane Science* 280 (1–2). Elsevier: 762–70.

Lu, Guang Yuan, and Wen Xiong Wang. 2019. "Water analysis | Seawater: Inorganic compounds for environmental analysis." In *Encyclopedia of Analytical Science*, 353–58. Elsevier.

Luo, Wenhai, Benedicta Arhatari, Stephen R. Gray, and Ming Xie. 2018. "Seeing is believing: insights from synchrotron infrared mapping for membrane fouling in osmotic membrane bioreactors." *Water Research* 137 (June). Elsevier Ltd: 355–61.

Madaeni, S. S., and Y. Mansourpanah. 2004. "Chemical cleaning of reverse osmosis membranes fouled by Whey." *Desalination* 161 (1). Elsevier: 13–24.

Malaisamy, Ramamoorthy, David Berry, Diane Holder, Lutgarde Raskin, Lori Lepak, and Kimberly L. Jones. 2010. "Development of reactive thin film polymer brush membranes to prevent biofouling." *Journal of Membrane Science* 350 (1–2). Elsevier: 361–70.

Matin, Asif, Z. Khan, S. M. J. Zaidi, and M. C. Boyce. 2011. "Biofouling in reverse osmosis membranes for seawater desalination: Phenomena and prevention." *Desalination*, 281: 1–16. Elsevier.

Miller, Daniel J., Paula A. Araújo, Patricia B. Correia, Matthew M. Ramsey, Joop C. Kruithof, Mark C.M. van Loosdrecht, Benny D. Freeman, Donald R. Paul, Marvin Whiteley, and Johannes S. Vrouwenvelder. 2012. "Short-term adhesion and long-term biofouling testing of polydopamine and poly(ethylene glycol) surface modifications of membranes and feed spacers for biofouling control." *Water Research* 46 (12). Elsevier Ltd: 3737–53.

Miura, Yuki, Yoshimasa Watanabe, and Satoshi Okabe. 2007. "Membrane Biofouling in Pilot-Scale Membrane Bioreactors (MBRs) treating municipal wastewater: Impact of biofilm formation." *Environmental Science and Technology* 41 (2). American Chemical Society : 632–38.

Moueddeb, H., P, J. P Jaouen, and F. Schlumpf. 1996. "Basis and limits of the fouling index: Application to the study of the fouling powder of nonsolubles substances in sea water." In *67–71.7th World Filtration Congress*. Budapest.

Naidu, Gayathri, Sanghyun Jeong, Saravanamuthu Vigneswaran, and Scott A. Rice. 2013. "Microbial activity in biofilter used as a pretreatment for seawater desalination." *Desalination* 309 (January). Elsevier: 254–60.

Ng, How Y., and Menachem Elimelech. 2004. "Influence of colloidal fouling on rejection of trace organic contaminants by reverse osmosis." *Journal of Membrane Science* 244 (1–2). Elsevier: 215–26.

Onoda, S. 2016. "Development of a novel wastewater treatment system combined direct up-concentration using forward osmosis membrane and anaerobic membrane bioreactor." Kobe University.

Parker, Jason A., and Jeannie L. Darby. 1995. "Particle-associated coliform in secondary effluents: Shielding from ultraviolet light disinfection." *Water Environment Research* 67 (7). Wiley: 1065–75.

Pearce, G. K. 2007. "The case for UF/MF pretreatment to RO in seawater applications." *Desalination* 203 (1–3). Elsevier: 286–95.

Richards, Melanie, and Thomas Eugene Cloete. 2010. "Nanozymes for biofilm removal." In *Nanotechnology in Water Treatment Applications*, 1–196. Edited by T. Eugene Cloete, Michele de Kwaadsteniet, Marelize Botes and J. Manuel López-Romero. Norfolk, UK: Caister Academic press.

Ridgway, H. F. 2003. *Biological Fouling of Separation Membranes Used in Water Treatment Applications*. USA: AWWA Research Foundation.

Sachit, Dawood Eisa, and John N. Veenstra. 2014. "Analysis of reverse osmosis membrane performance during desalination of simulated brackish surface waters." *Journal of Membrane Science* 453 (March). Elsevier: 136–54.

Sadr Ghayeni, S. B., P. J. Beatson, R. P. Schneider, and A. G. Fane. 1998. "Adhesion of waste water bacteria to reverse osmosis membranes." *Journal of Membrane Science* 138 (1). Elsevier Sci B.V.: 29–42.

Schneider, R. P., L. M. Ferreira, P. Binder, E. M. Bejarano, K. P. Góes, E. Slongo, C. R. Machado, and G. M. Z. Rosa. 2005. "Dynamics of organic carbon and of bacterial populations in a conventional pretreatment train of a reverse osmosis unit experiencing severe biofouling." *Journal of Membrane Science* 266 (1–2). Elsevier: 18–29.

Schwartz, Thomas, S. Hoffmann, and U. Obst. 2003. "Formation of natural biofilms during chlorine dioxide and u.v. Disinfection in a public drinking water distribution system." *Journal of Applied Microbiology* 95 (3):591–601.

Servais, Pierre, Gilles Billen, and Marie Claude Hascoët. 1987. "Determination of the biodegradable fraction of dissolved organic matter in waters." *Water Research* 21 (4). Pergamon: 445-50.

Shannon, Mark A., Paul W. Bohn, Menachem Elimelech, John G. Georgiadis, Benito J. Marīas, and Anne M. Mayes. 2008. "Science and technology for water purification in the coming decades." *Nature*, 452, 301–310. Nature Publishing Group.

She, Qianhong, Rong Wang, Anthony G. Fane, and Chuyang Y. Tang. 2016. "Membrane fouling in osmotically driven membrane processes: A review." *Journal of Membrane Science*, 499: 201–233. Elsevier.

Subramani, Arun, and Eric M. V. Hoek. 2010. "Biofilm formation, cleaning, reformation on polyamide composite membranes." *Desalination* 257 (1–3). Elsevier: 73–9.

Sun, Yan, Jiayu Tian, Liming Song, Shanshan Gao, Wenxin Shi, and Fuyi Cui. 2018. "Dynamic changes of the fouling layer in forward osmosis based membrane processes for municipal wastewater treatment." *Journal of Membrane Science* 549 (March). Elsevier B.V.: 523–32.

Van der Bruggen, B., J. Schaep, D. Wilms, and C. Vandecasteele. 1999. "Influence of molecular size, polarity and charge on the retention of organic molecules by nanofiltration." *Journal of Membrane Science* 156 (1). Elsevier Science Publishers B.V.: 29–41.

van der Kooij, Dick, Harm R. Veenendaal, Cynthia Baars-Lorist, Daan W. van der Klift, and Yvonne C. Drost. 1995. "Biofilm formation on surfaces of glass and teflon exposed to treated water." *Water Research* 29 (7). Pergamon: 1655–62.

van der Kooij, Dirk. 1992. "Assimilable organic carbon as an indicator of bacterial regrowth." *Journal of American Water Works Association* 84 (2): 57–65.

van der Kooij, Dirk, A. Visser, and W. A. M. Hijnen. 1982. "Dertermining the concentration of easily assimilable organic carbon in drinking water." *Journal of American Water Works Association* 74 (10). John Wiley & Sons, Ltd: 540–45.

Villacorte, Loreen O., Maria D. Kennedy, Gary L. Amy, and Jan C. Schippers. 2009. "The fate of transparent exopolymer particles (TEP) in integrated membrane systems: Removal through pre-treatment processes and deposition on reverse osmosis membranes." *Water Research* 43 (20). Elsevier Ltd: 5039–52.

Vrouwenvelder, J. S., and D. Van Der Kooij. 2001. "Diagnosis, prediction and prevention of biofouling of NF and RO membranes." *Desalination* 139 (1–3). Elsevier: 65–71.

Vrouwenvelder, J. S., D. A. Graf von der Schulenburg, J. C. Kruithof, M. L. Johns, and M. C. M. van Loosdrecht. 2009. "Biofouling of spiral-wound nanofiltration and reverse osmosis membranes: A feed spacer problem." *Water Research* 43 (3). Elsevier Ltd: 583–94.

Vrouwenvelder, J. S., J. W. N. M. Kappelhof, S. G. J. Heijman, J. C. Schippers, and D. van der Kooij. 2003. "Tools for fouling diagnosis of NF and RO membranes and assessment of the fouling potential of feed water." *Desalination* 157 (1–3). Elsevier: 361–65.

Vrouwenvelder, J. S., M. C. M. van Loosdrecht, and J. C. Kruithof. 2011. "Early warning of biofouling in spiral wound nanofiltration and reverse osmosis membranes." *Desalination* 265 (1–3). Elsevier: 206–12.

Vrouwenvelder, J. S., S. A. Manolarakis, H. R. Veenendaal, and D. Van Der Kooij. 2000. "Biofouling potential of chemicals used for scale control in RO and NF membranes." *Desalination* 132 (1–3). Elsevier: 1–10.

Vrouwenvelder, J. S., S. A. Manolarakis, J. P. van der Hoek, J. A. M. van Paassen, W. G. J. van der Meer, J. M. C. van Agtmaal, H. D. M. Prummel, J. C. Kruithof, and M. C. M. van Loosdrecht. 2008. "Quantitative biofouling diagnosis in full scale nanofiltration and reverse osmosis installations." *Water Research* 42 (19). Elsevier Ltd: 4856–68.

Vrouwenvelder, J. S., S. M. Bakker, L. P. Wessels, and J. A. M. van Paassen. 2007. "The membrane fouling simulator as a new tool for biofouling control of spiral-wound membranes." *Desalination* 204 (1–3 SPEC. ISS.). Elsevier: 170–74.

Wang, Jack Z., R. Scott Summers, and Richard J. Miltner. 1995. "Biofiltration performance: Part 1, relationship to biomass." *Journal - American Water Works Association* 87 (12). American Water Works Assoc: 55–63.

Wang, Xinhua, Yao Chen, Bo Yuan, Xiufen Li, and Yueping Ren. 2014. "Impacts of sludge retention time on sludge characteristics and membrane fouling in a submerged osmotic membrane bioreactor." *Bioresource Technology* 161 (June). Elsevier Ltd: 340–47.

Wang, Zhiwei, Junjian Zheng, Jixu Tang, Xinhua Wang, and Zhichao Wu. 2016. "A pilot-scale forward osmosis membrane system for concentrating low-strength municipal wastewater: Performance and implications." *Scientific Reports* 6 (1). Nature Publishing Group: 1–11.

Weinrich, Lauren A., Orren D. Schneider, and Mark W. LeChevallier. 2011. "Bioluminescence-based method for measuring assimilable organic carbon in pretreatment water for reverse osmosis membrane desalination." *Applied and Environmental Microbiology* 77 (3). American Society for Microbiology: 1148–50.

Wilbert, Michelle Chapman, John Pellegrino, and Andrew Zydney. 1998. "Bench-scale testing of surfactant-modified reverse osmosis/nanofiltration membranes." *Desalination* 115 (1). Elsevier Sci B.V.: 15–32.

Wingender, Jost, Thomas R. Neu, and Hans-Curt Flemming. 1999. Microbial extracellular polymeric substances : Characterization, structure and function. *Microbial Extracellular Polymeric Substances*, 1: XIV, 258. Springer Berlin Heidelberg.

3

Biofouling in Oil and Gas/Biofuel Pipelines

Dawn S S, Nirmala N, and Vinita Vishwakarma

CONTENTS

3.1 Introduction

Oil and gas and most recently biofuels are the most important industrial sectors, which govern the functioning of other industries across the globe. Transportation of this highly demanded energy resource has been a challenge, as even nations without oil and gas resources must rely on them creating a lot of import need; thus, they also indirectly influence the economic growth of the nation. The rising awareness on climate change and environmental and health impacts of pollution caused by the usage of conventional energy resources has now been included to the processing and transportation of biofuels too. The oil/gas/biofuel transported needs to be stored in designed containers to withstand the prevailing operating and environmental conditions. Materials used in the fabrication of transporting lines, vessels, or storage containers are to be carefully chosen owing to several material deterioration factors including the oil/gas/biofuels. Biofouling is one such

issue that causes the occurrence of cracks leading to leakage of fuel trans-
ported/stored in pipeline/storage tanks. Hence, it becomes a necessity to
understand the significance of these energy resources from both economic
and environmental points of view. An overview of the various causes of
biofouling, the mechanism by which biofouling occurs on materials, and its
impact and control measures have been discussed in detail.

3.2 Significance of Oil/Gas/Biofuel in the Energy Sector

The Global Industry Classification Standard (GICS) considers organizations
involved in the exploration, production, refining, storage, and transportation
of crude oil, natural gas, coal, and other fractions obtained on crude refining and alternate fuels including biodiesel, bioethanol, and biohydrogen to
be part of the energy sector (Frittelli et al. 2014). Petroleum has grown as an
essential product over the decades. As reported by the International Energy
Agency (2015), approximately 40% and 15% of the world's energy is available
from oil and gas. The importance of these commodities has reached so very
widely that a country's geopolitical and economic position is determined by
their availability, making other countries more dependent on them. Canada,
United States, and West European nations have been demonstrating to other
countries the measures that can be adopted to attain growth with economic
sustainability with energy source availability and longevity being the gov-
erning factors. Energy resources are inevitable in facilitating the economic
development of all other sectors in a country. Improper identification and
exploration of natural resources available in a nation thus lead to poverty
and deprivation, evident among many African and Middle East nations.
While natural resources have contributed to a great extent to determine the
economic growth of a country, they are witnessed to add equally toward the
prevalent corruption and unstable political scenario, and in the worst case
also led to the break of a warfare.

Countries wherein crude oil availability and production brings growth
in the economy, the governance and regulations concerned with petroleum
resources across the globe have gained importance. About 70% of a country's
Gross Domestic Product are contributed by oil and gas rentals. Proper governance of petroleum resources can create wealth for a country; hence, it may
be inferred that a proper administrative model will boost the performance
of the sector in those countries. Sustainability, self-reliance in energy, and
economic growth of other industries have been enabled through the administrative model adopted, which covers exploration and production of mineral
resources such as natural gas and oil which are the need for other industrial
sectors of the economy. The impacts of the oil and gas sector on a country's
economy irrespective of its developmental status are given in Table 3.1.

TABLE 3.1

Impacts of Oil and Gas Sector on a Nation's Economy

Countries Focused for the Study	Economic Impact	References
South Africa	It was interpreted that under the low and higher growth regimes, the oil price has predictive content for real GDP growth and sustainability, respectively.	Balcilar et al. (2017)
BRICS Countries (Brazil, Russia, India, China, and South Africa)	Panel data analysis was done to study the significance of energy consumed from 1990 to 2013 on the economic growth and reported that conversation, feedback, and neutrality hypothesis are valid for Russia, Brazil, and other countries, respectively.	Bayat, Tas, and Tasar (2017)
Fragile countries including India, Indonesia, Brazil, South Africa, and Turkey	Oil prices were correlated with GDP and current account deficit. GDP was reported to increase while the current account deficit decreased with oil prices.	Bayraktar (2016)
Russia	The economic growth was observed to be significant with no rise in the oil price.	Benedictow, Fjærtoft, and Løfsnæs (2013)
Organization of the Petroleum Exporting Countries (OPEC): United Arab Emirates, Kuwait, Saudi Arabia, and Venezuela	Varying oil prices and/or financial crisis influences the dependence of oil and economic growth.	Ftiti et al. (2016)
Australia	Using the multivariate generalized autoregressive conditional heteroscedasticity approach, dynamic movement of volatility in price returns of crude oil, coal, and natural gas was observed.	Ghassan and AlHajhoj (2016)
Iran	The effects of oil and non-oil exports on economic growth were studied and was found to have an inverse effect.	Parvin Hosseini and Tang (2014)
Cameroon	GDP, population growth rate, and petroleum rates were found to have a positive relationship with petroleum consumption. A contrary 'no casual relationship' was predicted between crude oil production and economic growth.	Sama and Tah (2016); Lecca et al. (2017)

Though the oil and gas sectors play a vital role in determining a nation's economy, the impact they have on changing climate has urged to understand the environmental impacts of oil and gas emphasizing a switchover partially or completely to biofuels. Table 3.2 gives the policy approaches implemented in various countries signifying the use of biofuels as alternates to oil and gas.

Be it petroleum-based fuel or biofuel obtained from various sources, one important issue of concern is the biofouling to which the transporting

TABLE 3.2

Policy Approaches in Countries Globally to Popularize Biofuel Usage

Countries	Policy Approaches
Brazil	27% and 10% blending of ethanol with gasoline and biodiesel with diesel to be imposed by 2019. Tax incentives were provided on ethanol and biodiesel-powered vehicles (Araújo 2018) on the basis of the available feedstock, size of production, and location to encourage alternate fuel popularization for alternate fuel commercialization. The compulsion to suppliers for the procurement of vegetable oil from family farms according to the National Biodiesel Production Program (PNPB) (Finco and Doppler 2011; Araújo 2018).
China	Subsidized grain-based biofuel programs were emphasized since 2000. The focus for producing 4 and 1 million tons of ethanol and biodiesel respectively was prioritized in order to enable a 10% biofuel mandate in vehicles by 2020.
European Union Countries	The 2009 European Union Energy and Climate Change Package (CCP) regulations emphasized 20% biofuels-based transport by 2020. Stringency in Green House Gas reductions, management of land as per the Renewable Energy Directive (RED), and implementation of non-food-based biofuels. Establishment of a 7% cap limits edible source-based first-generation alternate fuels by 2020.
India	Replacement of 20% petroleum-based fuel by 2017 and regulation of favorable conditions for biodiesel and bioethanol production including reduction of sugarcane prices. The national policies governing biofuels are framed to make India, a country that functions with a methanol economy with no import of petroleum from other major players across the globe.
United States	The objective of the Renewable Fuel Standards (RFS) is to achieve 36 billion gallons of biofuel consumption in 2022 from the nine billion gallons that existed in 2008 (Bracmort 2016).

pipelines and storage structures may be exposed owing to the hydrocarbon content in them and organic nature, which the microorganisms use as substrates. Hence, an in-depth understanding of the causes of biofouling, mechanisms, and control measures becomes essential in promoting the use of biofuels partially or completely.

3.3 Transportation Mechanisms

Barges, tankers, pipelines, trucks, and railroads are widely used for transporting crude oil from oil wells to refineries. More specifically, liquefied natural gas (LNG) and natural gas are transported in tankers and pipelines. Concerned with alternate fuels like ethanol and biodiesel, which are more corrosive than petroleum and diesel, transportation through pipelines is not recommended. Hence, trains, barges, and trucks are more popular for alternate fuel transportation though pipelines are more convenient. Additionally, ships are used for intercontinental transportation of biofuels. However, exhaustive studies are being pursued to modify the properties of ethanol and biodiesel, and their blends with petrol/ diesel are being pursued globally, so as to develop biofouling/corrosion-resistant pipelines to transport the ever-growing and demanded alternate fuels.

Until pipelines resisting biofouling and corrosion issues are commercialized, oil, gas, and biofuel industries will remain to use barges for their transport for long distances.

3.3.1 Oil Tankers

Oil tankers are defined as tank vessels that are designed and constructed exclusively for the transportation of oil and chemicals including hazardous materials in bulk. There are three major types of tankers used to transport crude oil; refined petroleum products like gasoline, jet fuel; black oils; and a variety of chemicals derived from crude petroleum. These tankers are designed to transport materials over long distances including sea routes. The major issue in using tankers is safety as stress on the hull is the concern.

Sagging, hogging, and shear force are identified to be the causes of the hull stress leading to crack development and leakage. Very Large Crude Carriers (VLCC) and Ultra Large Crude Carriers (ULCC) are the two configurations of supertankers available for carrying crude capable of transporting 2,000,000 barrels of oil.

3.3.2 LNG Tankers

Owing to high-pressure requirements, natural gas cannot be transported in oil carrying structures. Exclusive high-pressure withstanding structures are needed, making transportation a costly factor. Technologies have supported this need by giving options to transport natural gas in liquefied conditions, balancing both cost and transportation affordability. LNG obtained from natural gas processed at extremely low temperatures is transported in specially designed tankers with double hulls. This is an additional safety feature provided exclusively for LNG tankers to allow extra ballast as LNG is less dense than gasoline.

3.3.3 Pipelines

The Federal Energy Regulatory Commission (FERC) created in 1977 governs and regulates oil and gas transportation. Pipelines are a major transportation mechanism for the movement of oil. At least, part of the transportation of the oil occurs through pipelines. The crude oil separated from natural gas is transported through pipelines directly to a refinery. Gas transmission pipeline in the European Union countries is distributed in Germany, France, Italy, United Kingdom, Netherlands, Spain, Slovak Republic, Hungary, Belgium, Czech Republic, Austria, Greece, and Denmark. The length of pipeline in the United States for transporting natural gas is around 300,000 miles. The advancements in technology in material selection and monitoring systems have contributed to efficient and safe pipelines. Presently, the gas pipeline network existing in India is fabricated to cover 15,000 km. The government is laying exclusive schemes for the establishment of a gas grid network across the country. As per the planning, an additional 15,000 km gas pipeline has to be installed. Of which, the laying of pipelines covering 13,000 km distance has been authorized and the installations are in progress.

3.3.4 Barges

Barges are expensive with comparatively fewer infrastructure requirements than pipelines, primarily used on rivers and canals with less volume transporting capacity and more time-consuming loading ability. For transporting fuels across short distances by sea, barges are preferred; however, they are recommended only for calm waters.

3.3.5 Railroad/Tank Trucks

Railroads are an expensive mode of transportation when compared with pipelines. This mode of transport is more trusted for transportation of petroleum products from refineries to markets rather than transporting oil from wells to refineries. Gasoline and heating oils are transported to service stations and houses through tank trucks.

3.3.6 Tugboats

Mega ships are taken to ports with the aid of tugboats though toeing is the primary reason for which tugboats are preferred, they are also used in the transportation of oil and gas to transporting vessels.

3.4 Materials Used in Pipelines

Pipelines are majorly constructed using carbon steel owing to its wide availability, competent mechanical properties, and capability of acquiring higher resistance to corrosion by coating, lining, cladding, and chemical inhibition. Water content, presence of gases, pressure, temperature, flow rate, pH, and design life are the most important factors to be governed while selecting material for transportation of oil and gas. The increase in water content of the transporting material, be it oil, gas, or biofuel, alters the pipeline surface from being wetted by oil to wetted by water, making the pipeline more prone to corrosion. Oxygen, hydrogen, CO_2, and hydrogen sulfide present in the oil and gas transported induce corrosion and hence need to be eliminated if present. Hydrogen sulfide and hydrogen are the major sources causing Hydrogen Induced Corrosion (HIC), Sulfide Stress Cracking (SSC), and Stress-Oriented Hydrogen Induced Cracking (SOHIC). The toxicity and fatal characteristics even at low concentrations make hydrogen sulfide transportation risky owing to the associated health hazards. With the many disadvantages in transporting hydrogen sulfide, the only advantageous phenomenon is the formation of a protective ion sulfide film on the metallic substrates through which it is transported. The need for pipeline transport for the energy sector has resulted in the identification, selection, and popularization of non-metallic materials either used as coatings or standalones. When such materials are chosen for the construction of pipelines, pressure, temperature, and soil and sand movements when laid in deserts must be carefully considered.

With respect to handling biofuels, apart from considering the transportation pipelines, components of the engines also need to be critically reviewed for corrosion effects with the flow of biofuels in the engines. Especially with biodiesel, its corrosive nature toward the normal diesel engine and the fuel components is the major issue related to its usage in the normal engines. Conventionally, diesel engine components are made of copper and its alloys, aluminum and its alloys, cast iron, carbon steel, and stainless steel. Thus, a detailed review has been made in order to study the corrosion effects of aluminum and its alloys used in diesel engine components exposed to different types of biodiesel evolved from different sources and is given in Table 3.3.

Some of the corrosion-related properties determined in the Centre for Waste Management, Biofuel Laboratory of Sathyabama Institute of Science and Technology are tabulated as shown in Table 3.4.

Though biodiesel usage results in lowering the pollution emissions, the applications of biofuel have many challenges to be faced like the corrosion, tribo-corrosion phenomena, and on the other side the lower stability of fuel when in contact with metallic surfaces of the fuel transported system. Corrosion is generally caused due to the presence of metals in fuels, which causes abrasion. Due to this, the manufacturers are facing a lot of issues in the manufacturing of the mechanical components of the fuel system. This includes the static components like fuel tank HP fuel line fuel pipes injectors and also moving components like valves piston combustion chamber piston rings suffer from corrosion. Though the properties of biodiesel are similar to those of normal diesel, the corrosion aspects when used in the engine have to be studied. The non-presence of sulfur in the biodiesel reduces the corrosion

TABLE 3.3

Corrosion Rate of Aluminum and Its Alloys Exposed to Biodiesel and Blends Produced from Different Sources

Biodiesel Type	Experimental Conditions		Corrosion Rate in Millimeter per Year (mpy)
	Temperature (°C)	Duration (hours)	Aluminum and Its Alloys
Palm Biodiesel	RT	720	0.123
	RT	1,440	0.0527
	100	540	0.0002
Palm Biodiesel-Diesel-Ethanol (BDE)			
B20D75E5	RT	800	0.0614
B20D70E10	RT	800	0.0681
B20D75E5	60	400	0.212
B20D70E10	60	400	0.218
Pongamia pinnata biodiesel (B100)	RT	100	0.12 (static)
	RT	100	1.04 (flow)
	RT	100	1.04 (aerated)
	RT	100	0.16 (deaerated)
Pongamia pinnata B99 (1% NaCl)	RT	100	1.31 (static)
	RT	100	2.43 (flow)
	RT	100	2.43 (aerated)
	RT	100	1.19 (deaerated)
Jatropha curcas Biodiesel	15–40	7,200	0.0117
Karanja Biodiesel	15–40	7,200	0.0058
Mahua Biodiesel	15–40	7,200	0.0058
Salvadora Biodiesel	15–40	7,200	0.01236
Rapeseed oil Biodiesel	43	1,440	0.133
Vegetable oil deodorizer distillate biodiesel	45	2,880	0.0393

RT, room temperature.

in the fuel containers, but the main catalyst used in the trans-esterification process is acid or alkali-induced corrosion.

Overcoming this usage of the solid catalyst is better, so it can be easily separated from biodiesel. Purity is also the main concern in the biodiesel but the presence of glycerol, fatty acids, and alcohol, after the transesterification, has great effects like deposit formation, corrosion, and fuel system failure. The oxygen content in the biodiesel is the major cause of corrosion when used as a fuel. So, the usage of the corrosion inhibitors in diesel engines when the alternate biodiesel is used should be noted.

TABLE 3.4

Corrosion Related Parameters of Biodiesel Produced and Tested in Centre for Waste Management, Sathyabama Institute of Science and Technology

S. No.	Test Samples	Acid Value (%) ASTM D 664	Moisture Content (ASTM D 1500)	Copper Strip Corrosion @ 60°C for 3 hours (ASTM D 130)	References
1	Lipid	-	0.086	1a (slightly tarnish)	Nirmala and Dawn (2020)
2	Algal oil Biodiesel	0.48	-	1a (slightly tarnish)	Nirmala, Dawn, and Harindra (2020)
3	Waste Cooking oil biodiesel	0.37	-	1a (slightly tarnish)	
4	AOBD10: WCOBD 90	0.39	-	1a (slightly tarnish)	
5	AOBD20: WCOBD 80	0.38	-	1a (slightly tarnish)	
6	AOBD50: WCOBD 50	0.42	-	1a (slightly tarnish)	
7	AOBD70: WCOBD 30	0.46	--	1a (slightly tarnish)	
8	Conventional Diesel	0.03		1a (slightly tarnish)	
9	100% Waste Cooking oil (B100)	-	0.020	1a (slightly tarnish)	Ranjan et al. (2018)
10	20% WCOBD 80PBD(B20)	--	0.010	1a (slightly tarnish)	
11	10% WCOBD90PBD (B10)	-	0.008	1a (slightly tarnish)	
12	100% WCOBD blended with 30 ppm MgO nanoparticles (B100W30A)		0.012	1a (slightly tarnish)	
13	20%WCOBD80%PBD blended with 30 ppm MgO nanoparticles (B20W30A)	-	0.008	1a (slightly tarnish)	
14	10% WCOBD 90% PBD blended with MgO nanoparticles (B10W30A)	-	0.006	1a (slightly tarnish)	
15	New Zealand origin sheep skin ester (NOSSE)	-	1.09	1a (slightly tarnish)	Jayaprabakar et al. (2019)
16	New Zealand origin sheep skin biodiesel (NOSSB)	-	0.038	1a (slightly tarnish)	
17	New Zealand origin sheep skin biodiesel 5%+95% PBD (NOSSB5)	-	0.012	1a (slightly tarnish)	

(Continued)

TABLE 3.4 (*Continued*)

Corrosion Related Parameters of Biodiesel Produced and Tested in Centre for Waste Management, Sathyabama Institute of Science and Technology

S. No.	Test Samples	Acid Value (%) ASTM D 664	Moisture Content (ASTM D 1500)	Copper Strip Corrosion @ 60°C for 3 hours (ASTM D 130)	References
18	New Zealand origin sheep skin biodiesel 10%+90% PBD (NOSSB10)	-	0.016	1a (slightly tarnish)	
19	New Zealand origin sheep skin biodiesel 15%+85% PBD (NOSSB15)	-	0.019	1a (slightly tarnish)	
20	New Zealand origin sheep skin biodiesel 20%+80% PBD (NOSSB20)	-	0.021	1a (slightly tarnish)	

3.5 Biofouling Causes and Impacts on Oil/ Gas/Biofuel Transporting Pipelines

Biofouling is the unwanted accumulation of microbes forming a slime on the structures resulting in deterioration of both the substrate (such as pipelines or tank materials) on which the phenomenon prevails and the material that is transported, be it conventional fuel or biofuel. Biofouling majorly prevails in cooling water systems. Prolonged experience in the oil refining and petrochemical industry shows that two mechanisms of organic fouling are more likely to occur: (i) an auto-oxidation mechanism, which may be evident in hydrocarbon-based crude oil or gas as well as oxygen enriched biofuels, and (ii) deposition of asphaltenes (Groysman 2017). Whatever be the mechanism of fouling, it happens in five steps, namely, initiation, diffusion of the foulant (microorganism in this case) to the surface, microbial attachment, removal, and aging. Formation of fouling takes a minimum of 5 seconds to even months together and is temperature dependent (Watkinson and Li 2009) as is evident from the investigations reported earlier (Müller-Steinhagen, Malayeri, and Watkinson 2011; Yang et al. 2012; Yang and Crittenden 2012). Aging is a phenomenon related to fouling irrespective of the mechanism. As the deposits turn cohesive, they harden and adhere strongly to their structure. During biofouling occurrence in pipelines, microorganisms oxidizing iron are involved in the process, and the presence of ferric ions (Fe^{3+}) creates a solid shell, which further hardens and becomes a strong adherent on the pipeline (Müller-Steinhagen, Malayeri, and Watkinson 2011; Groysman 2010).

Biofouling leads to deterioration of fuel quality during transportation and storage because of the solubilization of the contaminants (foulants) into the material transported or stored, or discharging contaminants like H_2S by Sulfate-Reducing Bacteria (SRB) (biofouling) at the floor of the storage tanks storing kerosene and diesel fuel. Also, during shutdowns, the surfaces of equipment and structures made of carbon steel are opened up coming in contact with the atmosphere resulting in rust which flakes off.

Microbiologically influenced (or induced) corrosion (MIC) occurs on both inner and external surfaces in contact with the soil of the tanks and pipelines containing crude oil, gas oil (diesel fuel), kerosene (jet fuel), and fuel oil. MIC is caused by exclusively available bacteria; among the many bacteria and fungi types available, the ones that decompose crude oil containing organic components are MIC-inducing bacteria. There are a few types of bacteria that can disintegrate corrosion-inhibiting additives, thereby reducing their effectiveness. MIC is so dominant that it cannot be resisted by any metal or alloy available. Hence, research is necessary for developing materials that can tolerate MIC. The identification of MIC is complicated as in the majority of the cases, it occurs in combination with other corrosion types.

MIC can be identified by the coverage of sludge/slime/biofilm, which adheres to the metal surface due to the extracellular polymeric substance (EPS) secreted by the microorganisms. EPS not only adheres the biofilm to the metal structures but also aids in trapping and concentration of nutrients from the environment and protects the microbial attachments from biocides and toxic substances. The slipperiness of the biofilm is also because of the EPS secreted by the organisms. The growth of the microorganisms is less or rather not governed by the type or nature of the metallic substrate. Crude oil storing tanks may develop a black biofilm layer at the bottom. Similar evidences of grey, black, and greenish tinges may also be observed in tanks storing kerosene and gas oil. The microorganisms contributing to corrosion/biofouling are classified as SRB, microorganisms producing acids, microorganisms oxidizing ferrous (Fe^{2+}) and manganese (Mn^{2+}) cations, slime-forming bacteria, and methane- (methanogens) and hydrogen-producing bacteria.

3.6 Biofouling in Pipelines: Treatment and Controlling Mechanisms

Mechanisms adopted for controlling biofouling in pipelines/metallic substrates exposed to oil, gas, and biofuels are given below:

a. Crude oil preparation:
   Operations including settling, draining, and filtration may be adopted in the field before pumping the crude oils, petroleum

products, fuels, and organic solvents into the pipelines for transportation. Some clayey material, surfactants will still remain in the
crude and may end up in stable emulsions and are separated from
the oil by injecting emulsifiers and separating by desalting.

b. Crude oil always chosen as a tube side fluid in heat exchangers:
    To make cleaning of the tubes easier, crude oil is always allowed
to flow in the tube side while heating it in shell and tube exchangers. Cleaning of the tubes at the occurrence of fouling inside tubes is
easier than at the outside.

c. Facilitate caustic dispersion in the crude:
    To improve the dispersion of caustic solution in the crude, the con-
centration of the solution should not be more than 1.5 wt%.

d. Transporting oil/fuels at a higher velocity:
    Settling of solid particles to the bottom or on the surfaces can be
prevented when the oil fuel is transported at high velocities, thus
eliminating stagnation in the low-velocity regions and facilitating
the prevention of fouling occurrence.

e. Chemical treatment and usage of antifoulants:

Antifoulants are chemical substances that are injected into desalted crude
oil, or heavy petroleum products (atmospheric bottom, coker feed), and in
hydroprocessing stream to diminish or eliminate fouling. They can present
one or several chemicals. For instance, amine phosphate ester detergents,
inorganic phosphorus-containing acids, and salts thereof are used as antifoulants in crude oils. Usually, antifoulants contain a complicated mixture of dispersants, corrosion inhibitors, metal deactivators, and inhibitors
of polymerization. Each component has its purpose. Dispersants prevent
smaller particles from agglomerating to form larger particles that deposit
more easily. They also prevent the small particles from being attracted to
already existing deposits. Corrosion inhibitors prevent contact between
the metal surface and corrosive components. Metal deactivators 'neutral-
ize' metals by complexing, thus reducing the catalytic activity of the metal
(namely, Fe and Cu) to initiate polymerization. Polymerization inhibitors
react with radical (R·) to form stable molecules (R–R) and thus interrupt-
ing the chain reaction, which leads to the formation of polymers. It is not
easy to select antifoulant and its dosage because its efficiency depends on
the composition of fouling and hydrocarbon stream, temperature, conditions of its formation, flow regime, traits of unit operation, and its history.
Before the choice of antifoulant type, careful physicochemical analysis of
feedstock and deposits should be made. This analysis can indicate deposit
constituents (organic, inorganic, or a combination) and possible mechanisms of fouling formation.

3.7 Biofouling: Current Challenges and Impact on the Global Market

To reduce fouling in crude oil, gas, petroleum products, or biofuels, antifouling agents are becoming popular. Amine phosphate ester detergents, inorganic phosphorus-containing acids, and salts have been attempted and were found to be successful antifoulants. Generally, antifoulants are not individual compounds, they are a heterogeneous combination of dispersing agent's corrosion-resisting compounds, metal deactivating substances, and polymerization controlling compounds, so that each component functions independently and contributes to the overall antifouling effect aimed. Agglomeration of smaller particles to form larger masses is preventive by the dispersants. The contact between metal substrates and corrosion-causing substances is prevented by the anticorrosive agents. The catalytic activity is neutralized by the metal deactivators. The polymerization reaction is interrupted by converting radicals to stable molecules with the use of polymerization inhibitors. The selection of antifoulants and their concentration is highly dependent on the material that is stored or transported in the temperature flow region and the unit operations performed. The physicochemical characterization of the material into which the antifoulant addition is recommended is to be well studied on priority. Such an investigation can indicate the constituents and mechanisms of the fouling.

A combination of preventive and countermeasures is adopted in an integrated manner that is dependent on time irrespective of long- or short-term exposures. The most suitable strategy adopted is the choice of less adhesive materials, surfaces that can be easily cleaned, effective housekeeping practices, efficient and early warning systems like usage of sensors, use of limited nutrients, adoption of methodological cleaning practices, and monitoring measures. However, with all measures adopted, it is important to understand that there are enormous microorganisms that can cause biofouling and keep changing with time. Hence, it is important to learn to live with biofilms.

References

Araújo, Kathleen. 2018. *Low Carbon Energy Transitions.* Oxford University Press, 400 pages, ISBN: 9780199362554.

Balcilar, Mehmet, Reneé van Eyden, Josine Uwilingiye, and Rangan Gupta. 2017. "The impact of oil price on South African GDP growth: A bayesian markov switching-VAR analysis." *African Development Review* 29 (2). Blackwell Publishing Ltd: 319–36.

Bayat, Tayfur, Sebnem Tas, and Izzet Tasar. 2017. "Energy consumption is a deter-
minant of economic growth in BRICS countries or not?" *Asian Economic and
Financial Review* 7 (8). Asian Economic and Social Society: 823–35.

Bayraktar, Yuksel. 2016. "A casual relationship between oil prices current account
deficit, and economic growth an empirical analysis from fragile five countries."
Ecoforum Journal 5 (3): 29–44. http://www.ecoforumjournal.ro/index.php/eco/
article/view/474.

Benedictow, Andreas, Daniel Fjærtoft, and Ole Løfsnæs. 2013. "Oil dependency of the
russian economy: An econometric analysis." *Economic Modelling* 32 (1). North-
Holland: 400–28.

Bracmort, Kelsi. 2016. "The renewable fuel standard (RFS): Waiver authority and
modification of volumes," 10 pages, Library of Congress. Congressional
Research Service, Washington D.C. https://digital.library.unt.edu/ark:/67531/
metadc824757/.

Finco, Marcos Vinicius Alves, and Werner Doppler. 2011. "The Brazilian biodiesel
program and regional development: Cases from Northern Brazil." *Revista Do
Desenvolvimento Regional* 16 (3): 215–41.

Frittelli, John, Paul W. Parfomak, Jonathan L. Ramseur, Anthony Andrews, Robert
Firog, and Michael Ratner. 2014. *U.S. Rail Transportation of Crude Oil: Background
and Issues for Congress Specialist in Transportation Policy Specialist in Energy and
Infrastructure Policy.* USA: Library of Congress. Congressional Research Service.
www.crs.gov.

Ftiti, Zied, Khaled Guesmi, Frédéric Teulon, and Slim Chouachi. 2016. "Relationship
between crude oil prices and economic growth in selected OPEC countries."
Journal of Applied Business Research 32 (1). CIBER Institute: 11–22.

Ghassan, Hassan Belkacem, and Hassan Rafdan AlHajhoj. 2016. "Long run dynamic
volatilities between OPEC and Non-OPEC crude oil prices." *Applied Energy* 169
(May). Elsevier Ltd: 384–94.

Groysman, A. 2010. *Corrosion for Everybody.* Netherlands: Springer.

Groysman, Alec. 2017. *Corrosion Problems and Solutions in Oil Refining and Petrochemical
Industry.* Vol. 32. *Topics in Safety, Risk, Reliability and Quality.* Cham: Springer
International Publishing.

Jayaprabakar, J., S. S. Dawn, A. Ranjan, P. Priyadharsini, R. J. George, S. Sadaf, and
C. Rajeswara Rajha. 2019. "Process optimization for biodiesel production from
sheep skin and its performance, emission and combustion characterization in
CI engine." *Energy* 174 (May). Elsevier Ltd: 54–68.

Lecca, Patrizio, Peter G. McGregor, Kim J. Swales, and Marie Tamba. 2017. "The importance of learning for achieving the UK's targets for offshore wind." *Ecological
Economics* 135 (May). Elsevier B.V.: 259–68.

Müller-Steinhagen, H., M. R. Malayeri, and A. P. Watkinson. 2011. "Heat exchanger
fouling: Mitigation and cleaning strategies." *Heat Transfer Engineering*, 32:189–96.

Nirmala, N., and S. S. Dawn. 2020. "Phylogenetic analysis for identification of
lipid enriched microalgae and optimization of extraction conditions for biodiesel production using response surface methodology tool." *Biocatalysis and
Agricultural Biotechnology* 25 (May). Elsevier Ltd: 101603.

Nirmala, N., S. S. Dawn, and C. Harindra. 2020. "Analysis of performance and emission characteristics of waste cooking oil and chlorella variabilis MK039712.1
biodiesel blends in a single cylinder, four strokes diesel engine." *Renewable
Energy* 147 (March). Elsevier Ltd: 284–92.

Parvin Hosseini, Seyed Mehrshad, and Chor Foon Tang. 2014. "The effects of oil and non-oil exports on economic growth: A case study of the iranian economy." *Economic Research-Ekonomska Istraživanja* 27 (1). Taylor and Francis Ltd.: 427–41.

Ranjan, Alok, S. S. Dawn, J. Jayaprabakar, N. Nirmala, K. Saikiran, and S. Sai Sriram. 2018. "Experimental investigation on effect of MgO nanoparticles on cold flow properties, performance, emission and combustion characteristics of waste cooking oil biodiesel." *Fuel* 220 (May). Elsevier Ltd: 780–91.

Sama, Molem Chirstopher, and Ndifor Roger Tah. 2016. "The effect of energy consumption on economic growth in cameroon." *Asian Economic and Financial Review* 6 (9). Asian Economic and Social Society: 510–21.

Watkinson, A. P., and Y-H. Li. 2009. "Fouling characteristics of a heavy vacuum gas oil in the presence of dissolved oxygen." *Internation Conference on Heat Exchanger Fouling and Cleaning VIII-2009* 2009:27–32.

Yang, Mengyan, Andrew Young, Amir Niyetkaliyev, and Barry Crittenden. 2012. "Modelling fouling induction periods." *International Journal of Thermal Sciences* 51 (1). Elsevier Masson: 175–83.

Yang, Mengyan, and Barry Crittenden. 2012. "Fouling thresholds in bare tubes and tubes fitted with inserts." *Applied Energy* 89 (1). Elsevier Ltd: 67–73.

4

Introduction to Surface Coatings

K. Gobi Saravanan, A. M. Kamalan Kirubaharan,
and Vinita Vishwakarma

CONTENTS

4.1 Introduction

Cooling water pipelines significantly influence the energy conversion of thermal power plants and chemical plants. The cooling systems are classified as wet and dry systems. In the wet type, liquids have been used as a heat transfer medium and air has been used in dry systems. In general, water cooling systems are used in many thermal power plants due to higher heat transfer efficiency than air. In water cooling channels, liquids such as river water (fresh water) and sea water (sea water) are used as a heat transfer medium, which consumes a higher quantity of water. For instance, ash land water technologies is a leading chemical supplier company stated that around 3 million gallons of water has been used in the plants every day. The abundant quantity of water may comprise microbes, minerals, and other salts that lead to the formation of fouling in cooling systems. Alternatively, the use of pure water would not prevent bacterial adherence due to the presence of trace amounts of organic carbon, which favors biofouling. During bacterial growth in pipelines, extracellular polymeric substances (EPS) are secreted and accumulated onto solid surfaces that are eventually embedded in a highly hydrated environment (Christensen 1999). The secreted polymeric substances are mainly composed of proteins, polysaccharides, lipids, and nucleic acids (Flemming 2002). The EPS production will become a platform amidst bacterial cells and a solid material surface favors a higher mechanical stability of biofilms. Subsequently, the other organisms such as diatoms and microalgae will adhere onto the biofilm, which further facilitates the biofilm to grow further by entrapping nutrients from bulk water.

Control of biofilm in any kind of industry-related pipelines is a vital component of the successful water treatment program (Ludensky 2003). A theoretical analysis reported that avoiding initial bacterial adherence suppresses primary colonization and subsequent fouling. However, practically it is very tedious to maintain the antifouling surface owing to the slight adherence of bacteria sooner or later to the frequent cleaning process. In addition, Codony, Morató, and Mas (2005) clearly stated that intermittent chlorination treatment in the pipeline will tend to increase nearly tenfold and increase the bacterial load in the discharged water system. The routine monitoring protocols were highly assessed to identify the planktonic bacteria, whereas indigenous bacteria in marine aquatic systems still have major challenges to control. During the biofilm development, there are numerous circumstances such as pH of the medium, contact time, the concentration of liquids, surface chemistry of the solid surface, type of organic matters available in the surrounding medium, and the type of organisms present in the environment.

4.2 Surface Chemistry

The surface chemistry of the substratum determines bacterial accumulation and colonization. The less surface energy materials exhibited hydrophobic effects, which usually evolve less bacterial adhesion. The high surface energy materials show hydrophilic nature, which paves the way for higher bacterial adhesion. In many industries, the materials used to manufacture the pipelines are usually exhibited at a higher rougher surface. The higher rougher surface is susceptible to bacterial adhesion owing to having less bacterial size than crevices and less shear effect due to the irregular rougher surface. In addition, the composition of the material also plays a vital role in bacterial invasion. For instance, Lehtola et al. (2006) observed that *P. fluorescens* invasion is more pronounced on an aluminum plate rather than copper and brass and it was due to the antibacterial effect. Similarly, polyethylene pipes showed more bacterial colonies, whereas this effect was drastically reduced in copper due to its toxicity against microorganisms. Hence, the surface chemistry of the material also plays a significant role in bacterial invasion.

4.3 Substrate Preparation Techniques

Stainless steel is widely used for different applications ranging from dental implants, orthopedic implants, water, oil, and diesel transporting pipelines, storage tankers, etc. In addition, titanium material is also used for chemical and nuclear industries due to their better corrosion resistance, light weight, and mechanical properties. However, the high material cost of titanium compared to steel materials limits its wide angle of applications and has been used in specific components for industrial applications. Nowadays, surface coatings on stainless steel are an alternative way to improve their properties, thereby increasing the performance and durability of the materials. To make a surface modification, different methods are available, for this, substrate preparation and surface activation is an essential step to improve the adhesion, interfacial bond strength amidst the metal substrates, film uniformity, and crystallinity of the coatings. In the surface treatment process, different treatments such as acidic treatment, alkaline treatment, anodic pretreatment, and sand blasting are some of the important methods for effective surface treatment that are comprehensively used for metallic implant materials. During the pretreatment, soaking of the desired metallic surface in acidic and alkaline agents, H_2O_2 solution soaking, anodic oxidation, sand blasting process as well as the process with heating for the effective surface transformation process are done.

4.4 Surface Activation

Mild steel comprises low carbon (0.25%) and other trace elements such as silicon, manganese, sulfur, and phosphorous (Evans 1969). Mild steel is commonly used for almost all kinds of industrial applications due to its low cost and enough properties. However, atmospheric corrosion of mild steel is a major challenge to design engineers and requires suitable protection for its durability. Various types of surface coating such as thermal spraying, painting, powder coating, electroplating, plasma spray coating, and magnetron sputtering are the prerequisite methods to protect materials surface.

In general, the substrates are known to have 20–50 nm of native oxide on the surface of the substrates (Subramanian, Bhattacharya, and Memon 1995). However, no chemical treatment for oxide removal was carried out for Si substrates. The quartz substrate was used to study the optical properties of thin films (Kamalan Kirubaharan et al. 2015). Inconel-690 and Inconel-718 substrates in the size of 10 mm × 10 mm × 1 mm were cut from a thick square plate using a wire saw by electric discharge machining to maintain the uniformity at the cut region. The cut specimens were metallographically polished using silica emery grit paper starting from 220, 400, 600, 800, 1,000, 1,200, 1,500, 2,000 to 2,400. Finally, colloidal silica suspension (with a size of 1 μm) was used to reduce the surface roughness of the substrate. All the substrates were first cleaned in a soap solution to remove any contamination present on the substrate surface. This was followed by ultrasonic cleaning with water and acetone for 15 minutes and drying (Figure 4.1).

Broadly, two kinds of surface treatments are available: chemical and mechanical etching. The mechanical cleaning is further sub-divided into dry and wet cleaning. The chemical etching is to eliminate contaminants using emulsifiers and active materials in the cleaning solution (Mandich 2003). The materials that have undergone chemical treatment have several advantages such as a cleaned substrate need not be reassembled and dismantled, no damages occur on the material surface, irrespective of the sample size, and etching chemicals can be minimized depending on the size of the samples (Baumgärtner, Raub, and Gabe 1997).

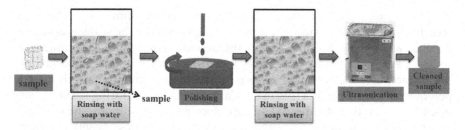

FIGURE 4.1
Schematic view of different stages of metal polishing.

4.4.1 Stainless Steel

Austenitic stainless steel (Grade 304 and 316L SS) constitutes chromium and other metals such as nickel, molybdenum, manganese, and iron. SS can form an oxide layer by corroding and create an iron oxide layer that shows very little protection in aggressive environments. Chromium presence in SS could improve the oxide layer and hence, the passive layer is formed about 1–10 nm thickness for the minimum concentrations of 10.5% Cr in the SS material. However, the passive behavior and corrosion protection against chemical attack are not as good as when passivated. During the manufacturing process, if the metal is not properly formed when alloyed with chromium in a correct composition, then rust is formed. To passivate the SS surface, acid treatment is generally used for the effective removal of manganese sulfide (MnS) or free ion inclusions present on the steel surface. The formation of MnS inclusions on the metal surface leads to defect points that propagate pitting corrosion thereby forming a discontinuous passive film that significantly deteriorates the SS surface.

The surface treatment influences the chromium in the SS to form chromium oxide (Cr_2O_3) known as a passive promoter owing to the strong metal-oxygen bond which is opposed to metalmetal bond strength favoring passive layer stability and facilitates nucleation sites on the metal surface for better oxide growth. The following parameters are conventionally available in industrial applications for the enhancement of materials durability.

4.4.2 Titanium and its Alloy

Titanium and its alloy materials that have been used widely for numerous applications possess good stability against chemical reactions and subsequently form a passive layer or oxide formation (TiO_2) during oxidation (2–6 nm thickness). The passive layer is a kind of surface barrier formed when exposed to air, water, and a hot environment. The thickness of the oxide layer may vary depending on the chemical and thermal treatment which limits the deterioration process. Furthermore, various surface treatments are available to enhance better adhesion and stronger bonds amidst the substrates and deposited films.

Surface preparation constitutes a surface treatment process, to modify the surface properties in a desirable way (Mattox 1996), and surface cleaning, which is the removal of surface contamination to a maximum extent. The main objective of any kind of surface preparation is to prepare the surface as homogeneous as possible. Modification of the desired sample surface comprises a series of surface treatment processes including roughening or smoothening the surface, creating a harder surface by plasma treatment, i.e., activating a polymer surface by plasma treatment. Care must be taken during surface treatment to ensure that the desired surface does not change in an uncontrolled and undesirable manner. Cleaning is an important process, where desirable film/coating properties can be obtained on the surface of a sample. Practically, the cleaning process is well known that contains no undesirable contaminants significantly. Hydrocarbons are common contaminants present on the surface layer that reduces nucleation density during deposition and diminishes the overall adhesion as well as the coating–substrate adhesion strength. Poor adhesion of film influences pin hole formation in the film surface which accelerates pitting corrosion. Furthermore, it facilitates insufficient electrical contact films in the case of electronic applications. The cleaning process must be economical and easy to meet the processing requirement. The rate of coating performance depends on the quality of the substrates. Surface preparation of the substrate is a prerequisite parameter before applying any kind of coatings. It is commonly known that appropriate surface preparation influences successful surface treatment. The presence of trace number of impurities such as grease, oil, and other oxide contaminants will seriously affect adhesion strength amidst coating and metallic substrates which should be minimized.

4.5 Alkaline Treatment

Alkaline pretreatment is commonly used to create a hydrated titanium oxide layer on the titanium surface. During the pretreatment process, hydroxide ions attack the titanium surface and subsequently form a sodium titanate hydrogel layer. During treatment, the formation of hydroxide group occurs as TiO_2 further dissolves in an alkaline environment. The formation of the hydroxide layer reacts with hydrated TiO_2 resulting in the formation of a more negative group which facilitates the sodium titanate layer. Since the layer was unstable, calcination was performed to enhance mechanical strength and crystalline nature. At elevated temperature, OH^- radicals present in the sodium titanate layer attack Ti surfaces which cause the transformation of highly crystalline titanate. Alkaline pretreatment creates an OH group on the metal surface. Once the materials are immersed in the alkaline medium, the oxide layer present on the metal surface dissolves and forms a metal hydride layer which creates a hydrous layer. After the alkaline treatment, the samples can then be exposed to a solution that contains Ni^+, Cu^+, Zn^+, and Mg^+ which will adsorb through ion exchange for the formation of Cu-P, Ni-P, and Mg-P coatings.

Alkaline treatment with thermal oxidation is an effective process to increase the thickness of the oxide layer. This technique has been carried out by keeping the substrate in the heating resistance furnace at a high temperature in the range of 400°C–1,000°C. Li et al. carried out alkali treatment using stainless steel with 10 M NaOH at 60°C for about 24 hours and subsequent calcination at 600°C for 1 hour. The hydrated phase transformed into sodium chromium oxide at 600°C and the phases were transformed to iron oxide and iron chromium oxide at 700°C–800°C phase. The formation of the iron oxide passive layer tends to decrease the strengthening of passivity leading to instability in the oxide layer and thereby peeling off the interface layer.

Heating the substrates above 600°C causes weakening of the passive layer obtained from loose structures of iron-chromium oxide and iron oxide, diminishing the adhesion strength between the material surface and coating. However, chromium oxide and sodium chromium oxide layer act as the initial protective layer on the steel surface during thermal alkali treatment.

Typically, alkaline salts such as sodium silicates, sodium phosphates, sodium carbonate, or sodium hydroxide are blended with wetting agents, surfactants, coupling agents, and solubilizers. The optimum temperature for alkaline treatment is 38°C–93°C. The composition of alkaline cleaners will vary depending on the size of the metal samples from which, how much quantity of impurities to be removed is determined. For stainless steel, aggressive (high ionic strength), high pH, and alkaline salts like potassium and sodium hydroxide salts can be used for effective cleaning. For aluminum, zinc, and brass cleaning, fewer pH solution builders such as potassium and sodium silicates, bicarbonate, and borate salts are predominantly used.

Various mechanisms such as emulsification, saponification, and conversion are generally used for cleaning the materials. Emulsification is a kind of process in which two insoluble compounds can be combined and removed by a surfactant. For example, the petroleum oil and water deposited onto solid materials could be removed by adding a surfactant which has a hydrophobic or oil-soluble end as well as a hydrophilic water-soluble end. Dispersion is the type of cleaning process which makes the surface wet thereby penetrating the oil film. At the metal–liquid interface, the surfactants will certainly reduce the interfacial and surface tension of the cleaning solution (especially oil). The cleaner undercuts and penetrates the oil and breaks their bonds resulting in the formation of tiny droplets, which finally float to the surface.

4.6 Acidic Pretreatment

Acidic pretreatment is an effective and efficient method to remove surface debris and native oxides on the material's surface. During acid treatment, MnS debris was removed substantially on the steel surface and formed a strong passive layer composed of chromium content as well as enrichment of noble elements present in the steel surface. The main advantage of the acid treatment is to develop stable and dense passive layer formation which protects the surface from a hostile environment. In this view, Kannan et al. conducted an acid pretreatment experiment on steel surfaces using different concentrations of sulfuric acid ranging from 5% to 20% at room temperature for 60 minutes. After acid treatment, the samples were gently washed with distilled water and then dried at 50°C. After acid treatment, the surface elemental composition was determined by elemental dispersive X-ray spectroscopy (EDX). The corrosion protection behavior of acid-treated samples was further analyzed by electrochemical cyclic polarization studies. After electrochemical studies, the ions leached from the surface were analyzed by inductively coupled plasma atomic emission spectroscopy (ICP-AES). Among the various concentrations of H_2SO_4 in acid treatment, 15% of H_2SO_4 was fixed for the electrochemical corrosion test. The acid-treated steel samples exhibited a higher break potential of +680 mV, which is almost double-fold increment than the value of standard pristine 316L SS (+320 mV) indicating superior corrosion resistance. In addition, polarization resistance (R_p) and electrical impedance behavior ($|Z|$) of acid-treated samples showed noble values (R_p was 126.2 and $|Z|$ was 2.09) at the concentration of 15% H_2SO_4 rather than untreated steel (R_p was 43 and $|Z|$ was 1.61). The corrosion protection behavior of acid-treated steel was exponentially improved due to the formation of chromium oxide along with Mo enrichment at the 316L SS surface. In order to prove this, EDX and ICP-AES were performed, and the results revealed that a higher quantity of Cr and Mo, as well as a lower quantity of Fe, was shown on the acid-treated steel surface. At the same time, fewer Cr and Mo elements and higher Fe elements were present while the steel surfaces were treated with 10% and 20% concentration of H_2SO_4. These studies strongly indicated that 15% of H_2SO_4 was an ideal concentration that significantly fights against pitting corrosion due to the integration of Mo and Cr in a harsh environment.

Other acid treatments such as nitric acid, phosphoric acid, and citric acid demonstrated similar effects on steel surfaces like the sulfuric acid treatment process. Noh et al. carried out the pretreatment on steel using different concentrations of nitric acid up to 50 wt% at room temperature for 60 minutes. From the results, they stated that MnS inclusions were effectively removed at 25 wt% of nitric acid treatment and subsequently formed chromium oxide enrichment when treated with 20–25 wt% of acid solution. Similarly, Hagiwara et al. (2015) examined the pretreatment effect on 316L SS using citric acid or a combination of citric acid and nitric acid solution. For this, citric acid (1.05%) was used to pretreat the steel surface for 120 minutes at 30°C–80°C. Then, pretreatment with nitric acid (1.05% or 4.55%) mixed with 1.05% of citric acid at 60°C which mimics the pasteurization conditions of egg products was carried out. From the study, they strongly concluded that the pretreatment time was effectively suppressed from 120 to 15 minutes, which was still highly effective for preventing protein adsorption. In addition, 1.05% nitric acid containing 1.05% citric acid could be an excellent choice for effectively removing adhered proteins of food manufacturing equipment that was fouled with egg products.

4.7 Types of Surface Cleaning

Removal of gross contamination from the desired surface is the initial step for the cleaning process. Various materials such as emery paper, sandpaper, steel wool, and scotch-brite are generally used. In addition, abrasive powders including Al_2O_3, SiC, precipitated calcium carbonate ($CaCO_3$), diamond, and CeO are significantly used.

4.7.1 Immersion Cleaning

The facile method for using an alkaline solution as a cleaner is by immersion. The part of the sample to be cleaned is placed on a rack or hook and dipped in the cleaner solution to confirm that all of the sample parts are below the cleaner solution. An appropriate quantity, concentration, temperature, duration, and pH of the solution for an alkaline cleaner would be ca 80 g/L at 80°C for 5 minutes. Surface cleaning of the solution by alkaline immersion is one of the least expensive methods in terms of equipment and lesser time duration for the overall process. The alkaline solution container must be made of a vessel to heat the solution for effective cleaning.

4.7.2 Pickling Process

Normally, pickling solution comprises a mixture of nitric acid and hydrofluoric acids (phosphoric acid can be used to obtain mild pickling properties), with surface-active and binding agents to obtain the right viscosity and good thixotropy. The method is generally used for large surfaces. Proper rinsing with clean tap water must be followed prior to pickling treatment. The quality of rinsing water should be acceptable and the chloride content should be increased according to the surface requirements. Pickling treatment normally involves a mixture of acidic solutions containing 8–20 vol% of nitric acid (HNO_3) and hydrofluoric acid (0.5–5 vol%). However, chloride-containing agents like hydrochloric acid (HCl) should be avoided or used in very less quantity due to the high risk of pitting corrosion.

After preparing a metal piece, the manufacturer or research personnel will submerge the samples into the acid solution. While incubating into pickle liquor, the acid solution will eat away any impurities or oxides present on the surface of the metal samples (Figure 4.2).

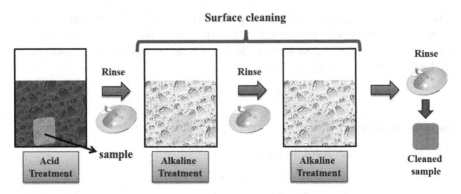

FIGURE 4.2
Schematic diagram of steps involved in surface cleaning by acid/alkaline solutions.

In addition, there are other ways available to clean and descale the metal. In order to fabricate a smooth surface, smooth cleaning is an alternative approach that offers similar effects in surface cleaning. With a smooth surface, metal is exhibited with an abrasive compound that removes surface impurities and imperfections physically. In addition, abrasive blasting is also used in which, blasting of the samples was carried out in a pressurized stream of an abrasive solution to fabricate a smooth and clean surface.

4.7.3 Pickling: Benefits

During manufacturing of the material, an oxide layer develops and alters surface roughness. The presence of this oxide layer diminishes clean and smooth surfaces. Though there are various methods to remove oxide and other impurities from steel, pickling is an effective treatment offered by most of the manufacturing companies owing to its simplicity. During pickling treatment, the pickle liquor removes oxide of surface steel without altering the chemical or physical nature of underlying steel.

Although metal does not show oxide on its surface, it still contains some impurities. It is not unusual for a newly manufactured metal to present inorganic compounds like trace elements. The manufactured metal is intended for different applications, where these impurities may obstruct its performance.

4.7.4 Electrocleaning

Electrocleaning is a specific technique of immersion cleaning. Electrocleaning is closely similar to immersion cleaners (similar composition), except that the surfactant concentration is usually lower; hence, it exhibits less foaming. The cleaning action of the metals and their alloys is assisted by using direct-current electricity. In the electrocleaning process, two electrodes are placed at two corners of the vessel which must be immersed into a cleaning solution, where the cathode electrode carries a negative charge. Particularly, the sample to be treated carries a positive charge (anode). During the cleaning process, the oxygen evolving at the sample surface acts as a mechanical scrubber supporting the removal of soil contaminants. The higher solution concentrations (75–125 g/L) and elevated temperature (77°C–99°C) are usually maintained in electrocleaning rather than straight immersion cleaning methods. Furthermore, a large rectifier is used for supplying direct current, and the current density is given in the range of 27–160 mA/cm². Electrocleaning is usually carried out by spray or immersion to remove soil adhered onto the sample surface. The cleaner sample is typically left in an electro cleaner bath for 1–3 minutes to acquire an exceptionally smooth and clean surface.

4.7.5 Electropolishing

When a metal and its alloys are immersed in a solution under adequate electrolysis conditions, it may involve anodic dissolution resulting in the surface to become smooth and glossy (Clerc, Datta, and Landolt 1984). Electropolishing (EP) is normally used to produce optimal corrosion resistance capability, surface roughness, and shining of steel surfaces. EP is a kind of technique, in which the current density, temperature, time, and pH of the bath solution play a pivotal role to obtain shiny and corrosion-resistant material surfaces (Figure 4.3). In a typical electrochemical polishing, the anodic and cathodic reactions are as follows:

$$M^\circ - ze^- \rightarrow M^{z+} - 1$$

$$2H^+ + 2e \rightarrow H_2 - 2$$

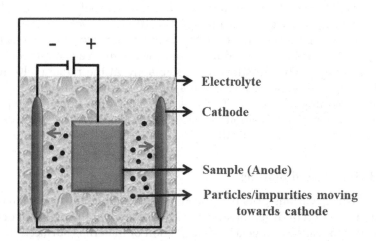

FIGURE 4.3
Schematic diagram of electropolishing.

4.7.6 Manual Polishing

In general, the polishing method comprises two types, namely, macropolishing and micropolishing. Macropolishing deals with the grinding of metal surfaces to the approximate size of 100 µm, whereas micropolishing removes the top surface to 10 µm approximately.

4.7.7 Barrel Cleaning

Barrel cleaning is a little variation of immersion cleaning where the samples are immersed in a six- or eight-sided barrel, of which, all the sides are perforated, allowing the cleaning solution to enter the barrel. The barrel is dipped into cleaner and then rotated around the long axis using an electric motor. The sample inside tumble and polish against each other, as a result aiding in the elimination of the soil. Electrodes are also inserted and connected to a rectifier to assist in effective electrocleaning. However, small parts such as cabinet handles, fasteners, and screwdriver shafts are only used in barrel cleaning.

4.7.8 Spray Cleaning

The other principal way of cleaning the sample is by spray. In spray cleaning, the solution is discharged from the holding tank into progressive pipes (risers), which constitutes other pipes known as headers coming out from the risers. These headers consist of holes drilled into them and nozzles inserted into the holes. The pipes are designed in a way that a part going by on a conveyor tube would be sprayed from every conceivable angle with a cleaner solution. This setup configuring a series of pipes is housed in a steel vessel, which aids the spray mist confined. The hood is normally over the holding tank facilitating the excess of cleaner solution draining back into the holding tank. The hood has a fine slit at the top to allow conveyor parts racks to go through and one exhaust is used to remove the steam. In a commercial spray cleaner technique, the pressures for systems are in the range of 10–40 psi (70–275 KPa), bath concentrations are 4–30 g/L characteristically, and bath temperatures range amidst 21°C–88°C. The most important difference in immersion cleaner composition is that the surfactant concentration is low, which reduces foaming drastically.

4.7.9 Grit and Sand Blasting

In grit blasting, various grits have been used including alumina, silica, and fractured cast iron of different sizes which accelerate in the presence of high gas steam that causes gouge and deformation of materials surface. The fine particles can be entertained with high velocity by a siphon system or pressure system, which is used in sand blasting equipment.

4.7.10 Wet Chemical Cleaning

Chemical etching is used to remove the surface material along with tiny contaminants. It is a very useful method for cleaning the surface into a proper condition. Pickling is a kind of process used to remove large amounts of surface oxides that are formed on metal surfaces during the metal manufacturing process. The process commonly underlies an alkaline cleaning prior to acid pickling to get uniform etching and wetting of surfaces. For instance, aluminum and its alloys can be treated by the pickling process by dipping with various combinations/concentrations of nitric, sulfuric, chromic, and hydrofluoric acids. Similarly, copper and its alloys are pickled in the presence of oxidizing and sulfuric acids with various concentrations. Steel and iron surfaces can be pickled in hydrochloric acid or sulfuric solutions. Chemical etchants are highly aggressive in their action and etch preferentially at grain boundaries in a two-phase system. Cleaning of metals by acid cleaning has the deleterious effect of hydrogen formation into the metal and ceramic surface (Cuthrell 1979).

The etching method particularly removes the oxide layer at the top of the surface. Common etchants for glass include ammonium bifluoride (dissolve 100 g of ammonium bifluoride in 800 mL of distilled water) or sodium, trisodium phosphate, which act as mild etchants; hydrofluoric (HF) acid acts as a very strong etchant. In addition, HF acid is a common and powerful etchant for silicon and metal surface, which can leave the silicon surface either hydrogen terminated or hydroxyl terminated.

To some extent, chemical etching could not remove some matters from a surface and leaves a smut, which should be properly removed by another etching process. For example, etching copper-containing aluminum alloys with NaOH leaves a copper smut and/or a silicon smut on the surface. The copper smut can be removed by an HNO_3 etch and a copper/silicon smut can be removed with an HNO_3/HF etch. In some cases, an etchant can be devised that etches all the constituents uniformly. For example, in etching Al:Cu:Si alloys, a concentrated nitric acid (100 mL) plus ammonium bifluoride (6.8 g) etch is used. During etching process, copper, aluminum and silicon will be etched on surface of Al:Cu:Si alloys resulting the removal of oxides with desired patterns. The etchant actually etches silicon more rapidly than aluminum.

4.8 Types of Coating Techniques

4.8.1 Electroless Plating

In the electroless nickel plating technique, the bare sample is dipped into a bath containing metal ion deposit on the samples by triggering the action of reducing agent. As per the difference in pH and temperature in the plating bath, the electroless plating can be classified into alkali electroless or acid electroless

FIGURE 4.4
Flow chart of the electroless plating process.

plating. In electroless plating, the bath solution constitutes respective salt solutions (nickel, copper, zinc, and so on), reducing agents (borohydride or sodium hypophosphite), stabilizing agents (thiourea, metal salts, and so on), a complexing agent (succinic acid, glycolic acid, lactic acid, citric acid, and acetic acid), and finally pH conditioning solutions such as hydrochloric acid and ammonia. In addition, multiple electroless plating can be done by admixing desired elements into the bath (Malecki and Micek-Ilnicka 2000). In general, conventional electroplating is usually conducted as given in Figure 4.4.

The technology has been improved by optimizing ideal process parameters including better adhesion, uniform coating, and surface smoothness of films. For this, different kinds of reducing or complexing agents, pH of the bath solution, plating, and post-plating temperature predominantly determine the quality of the coating.

4.8.2 Reducing Agents

Reducing agents play the most important role in electroless plating, reducing metal salt to metal ions without changing its oxide nature. For example, in the Cu electroless plating method, cupric ions (Cu^{2+}) are converted into Cu metal ions (Cu) without the formation of cuprous oxide (Cu^+). Traditionally, formaldehyde is often used as a reducing agent in the electroless copper plating method. Meerakker et al. used various reducing agents such as phenyl hydrazine, dimethylamine borane (DMAB), aminoborane, and hydrazine. From the study, they illustrated that the development of coating for commercial purposes will be challengeable due to deposition quality and bath stability optimizations (van den Meerakker and de Bakker 1990).

4.8.3 Complexing Bath

In an electroless copper bath, the complexing agents play a vital role in good-quality deposits. In general, complexing agents minimize the metal to metal hydroxide formation in the alkaline pH as well as increase the bath life. In addition, a mixture of complexing agents in small quantities enhances the plating rate. Trisodium citrate, lactic acid, ethylene diaminetetra acetic acid (EDTA), malic acid, sodium potassium tartarate, triethanolamine, etc. are the most common complexing agents used for electroless plating methods (Fuchs-Godec, Pavlovic, and Tomic 2015; PauliukaitĖ et al. 2006).

4.8.4 Additives

Additives are commonly used to prevent bath decomposition and improve the effective electroless coating process. They behave as bath stabilizers and improve the mechanical properties of metal deposits. Furthermore, additives are used in various roles in electroless plating.

Inhibitors: additives are used to throwing power into recess and holes. Ex. polyoxyethers and polyether.

4.8.5 Levelers

Levelers determine plating thickness and uniformity at the corner of the samples and level in the plated layers. Ex. amide and amine surfactants.

Brighteners: Brighteners especially alter plating rates and control microstructural characteristics such as grains and crystallinity of the deposited coatings. They also determine hardness and mechanical properties of the deposits.

4.8.6 Wetting Agents

Wetting agents like surfactants in the bath facilitate lower surface tension of the electroless bath solution, thus allowing a better wetting nature of the bath solution. The common additives for wetting agents are thiourea, cytosine, pyridine, guanine, glycine, adenine, ammonia, etc. Most importantly, the additive is mostly involved in the tuning of crystal size, orientation, and shape of the grains (Paunovic and Schlesinger 2006).

4.8.7 pH of the Bath Solution

Bath pH is a vital parameter in electroless plating. Bath pH influences plating rate, surface roughness, microstructure, and crystallinity of coated films (Duffy, Pearson, and Paunovic 1983). Reducing agents in the bath solution trigger the formation of hydroxyl ions (OH^-). However, pH should be stable in the bath solution for better efficiency in the plating process. If the change in pH occurs, severely affecting the deposition rate while plating, the mechanical properties get affected in the coating. To stabilize the pH values in the bath solution, various stabilizers such as KOH, NaOH, amines, and carboxylic acids are commonly used in alkaline solutions.

4.9 Binary Electroless Plating Process

Electroless deposition of binary palladium-phosphorous coating exhibited better properties rather than metal coating alone. For example, Ni-P/Au and Ni-P/Ag nanocomposite coating showed higher hardness. Furthermore, nanocomposite coatings achieved crystalline nature from amorphous nature at 335°C (Ma et al. 2009). Robertis et al. deposited P-Pd by electroless plating and investigated the mechanism of metal/solution interactions at equilibrium potential by electrochemical impedance spectroscopy. Ashassi et al. studied the impact of the complexing agent in structural and potentiodynamic polarization studies using the weight-loss method. The complexing agent sodium citrate enhanced the microhardness of Ni-P composite coatings with higher corrosion resistant nature in 3.5% NaCl solution (Ashassi-Sorkhabi and Es'haghi 2013).

4.10 Powder Spray Deposition

In the spray process, the pre-mixed powder is injected along with the carrier gas through a high-energy source. The mixed powder is sprayed in the form of molten or semi-molten droplets and directly adhered onto the components (Ning, Jang, and Kim 2007). The principle of this process is to feed the powder into the plasma where the particles are rapidly heated up to their melting point and accelerated to speeds in the order of 300 m/s. The molten powder particles stuck onto the substrate surface and solidified rapidly. The adhesion between the particles and the substrate is mainly through a mechanical bonding. The deposition rate and the quality of the coatings depend upon the type of spray method selected, the processing parameters, type of coating, and the substrates used (Behera and Mishra 2012). The major advantages of this process are: (i) relatively high deposition rate and (ii) the various metallic and oxide coatings that can be applied. A significant disadvantage of the spray processes is the inability to obtain homogeneous, high quality, and dense coatings (Ana and Maria 2016).

4.11 Chemical Vapor Deposition (CVD) Technique

Some of the difficulties in the spray process can be overcome by the CVD techniques. In the CVD process, a reactant gas mixture is passed in a high-temperature reactor to form a solid thin film on the substrate's surface. In general, the CVD coating process takes place at high temperatures (<1273 K). Various metal carbides, nitrides, and ceramic coatings have been deposited at the rates of 5–10 µm/hour (Mallika and Komanduri 1999). The disadvantages of the CVD process is that they often require high-deposition temperatures and it produces chemical waste (such as acids) that is environmentally unacceptable and the deposition rates are usually low for high-quality coatings (Pauleau 2002).

4.12 Physical Vapor Deposition (PVD) Technique

Some of the shortcomings of the CVD and spray processes can be addressed by the PVD methods. In the PVD processes, the coating material is evaporated by various methods (resistance heating, high-energy ionized gas bombardment, or electron gun) under vacuum conditions. The atoms in the vapor phase are transported by means of a line-of-sight process to the substrate to form a coating. The PVD coating process takes place between room temperature (RT) and about 973 K. The PVD technique is well-suited for new and advanced coating concepts like gradient coatings, multi-component coatings, and multilayer or superlattice coatings.

Sputtering is one of the most versatile PVD processes available for thin-film deposition. Various metallic and ceramic (carbides, oxides, and nitrides) coatings can be deposited by this process usually at a rate of a few μm or less per hour. Unlike the CVD process, PVD processes are clean, non-toxic, and environmentally friendly. The main disadvantages of PVD processes (with exception of EBPVD) are (i) relatively low deposition rates (0.1–0.6 nm/s) and difficulty in applying oxide coatings efficiently on all the sides of a component. In spite of significant advancements in the various PVD processes, there are still drawbacks in the coating quality. For example, the cathodic arc PVD process produces macro-particles of metals (1–15 μm in size) during evaporation. These molten particles produce non-homogeneity in the microstructure and produce unfavorable physical properties. Chemical and physical conditions during the deposition reaction can strongly affect the chemical composition, residual stresses, and microstructure (i.e., crystallinity and epitaxial) of the coatings. The effect of these conditions must be understood to control the process. Table 4.1 provides a brief comparison of various coating processes.

TABLE 4.1

Comparison Among PVD, CVD, Spray Process and EBPVD Techniques (Singh and Wolfe 2005)

Coating Process	Deposition Rate	Surface Roughness	Microstructure	Substrate Temperature
Powder Spray	>100 μm/minute	Very High	Lamellar	RT to 1,073 K
CVD	>5 μm/hour	Moderate	Columnar	>1,073–1,473 K
PVD – Sputtering	>5 μm/hour	Low	Columnar	Less than 973 K
EBPVD	100 nm to μm/hour	Low	Columnar	RT to 1,073 K

4.13 Electron Beam Physical Vapor Deposition Technique

Electron beam physical vapor deposition (EBPVD) is a derivative of the electron beam melting technique. The EBPVD process has overcome some of the complexities associated with the CVD, spraying, and other PVD processes. In the EBPVD process, high-energy electron beams generated from an electron gun are focused and directed to evaporate the material to be deposited. The evaporant materials travel in a vacuum and condense onto the substrates resulting in the formation of a continuous film. During deposition, substrate heating is often applied to enhance the metallurgical bonding (enhancing diffusion process for good adhesion) between the coating and the substrate. The EBPVD process is a line-of-sight process, and therefore, uniform coating of complex parts can be accomplished by rotating the substrate holder in the vapor cloud during the deposition process (Figure 4.5).

There are five main components in the EBPVD system, namely, gun assembly, water-cooled copper hearth, substrate holder, power source, and the vacuum chamber. The schematic diagram of EBPVD is shown in Figure 4.5. The electron beam can be accelerated by an applied potential and can pass through the magnetic field where it gets deflected through 270°C, and it strikes the evaporant material placed in a water-cooled copper hearth (Figure 4.5).

The process parameters such as e-beam power, chamber pressure, substrate–source distance, substrate rotation, and substrate temperature affect the final coating structures.

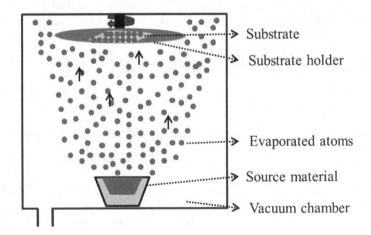

Substrate

Substrate holder

Evaporated atoms

Source material

Vacuum chamber

FIGURE 4.5
Schematic diagram of the EBPVD process.

4.14 Thin-Film Growth Characteristics

The atomic species produced in the PVD processes by electron beam, plasma, or laser are transported to the substrate through vacuum and finally condense on the substrate surface forming a thin film (SreeHarsha 2006). In general, the films will be deposited at low pressures to avoid the reaction between the depositing species and the chamber atmosphere. At these low pressures, the mean free path of the depositing species is high, and they travel in straight lines toward the substrate surface (line-of-sight process) (SreeHarsha 2006).

Thin-film growth is a complex process, where a number of steps are involved at the nanoscale level. They are the arrival of adatom, nucleation, island growth, coalescence of islands, formation of polycrystalline islands and channels, development of a continuous structure, and film growth. The detailed step-by-step growth process based on various experimental and theoretical studies (Petrov et al. 2003) are summarized in Figure 4.6.

The adatoms arrive on the substrate surface, losing their velocity component and are physically adsorbed onto the substrate surface. Initially, the adsorbed adatoms are not in thermal equilibrium with the substrate and moveover the substrate surface. In this process, they colloid themselves leading to the formation of bigger clusters. If the clusters are thermodynamically

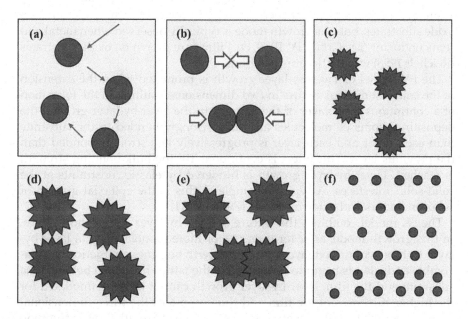

FIGURE 4.6
Thin film growth kinetics. (a) Arrival of single atom. (b) Doublet formation. (c) Nucleation. (d) Growth. (e) Coalescence. (f) Continuous film.

unstable, they may tend to desorb from the substrate surface depending on the deposition parameters. The adsorbed clusters collide with other adsorbed species and start growing in size, which is termed as the nucleation process. After reaching a certain critical size, the cluster becomes thermodynamically stable and the nucleation barrier is said to have been overcome. The nuclei grow in number on the substrate surface until a saturation nucleation density is reached. The nucleation barrier depends on the kinetic energy of the adatoms, the rate of impingement, the activation energies of adsorption, desorption, thermal diffusion, substrate temperature, and the chemical nature of the substrate. A nucleus can grow on the substrate by diffusion of the adsorbed species to form islands. The next stage of film formation is the island coalescence, in which the small islands start coalescing each other. Larger islands grow together, leaving channels and holes of uncovered substrate. At this stage, the film structure changes from discontinuous island type to a porous film structure. Filled-up channels and holes result in the formation of a continuous film.

The thin-film growth model can be categorized into (i) island growth, called Volmer–Weber (VW) model; (ii) layer-by-layer growth, called Frank–Van der Merwe (FV) model; and (iii) layer plus island growth, called Stranski–Krastanov (SK) model (Ranjan 2005).

In the VW model or island growth, the atoms or molecules being deposited are more strongly bonded to each other than to the substrate material. This is likely to happen when the film and substrate are dissimilar materials. There are a few examples of such behavior in the growth of oxide films on oxide substrates, but this growth mode is typically observed when metal and semiconductor (i.e., Group IV, III–V, etc.) films are grown on oxide substrates (Reichelt 1988) with lattice mismatching.

The FV model or layer-by-layer growth is promoted when the extension of the smallest nucleus occurs in two dimensions resulting in the formation of a continuous monolayer of the deposit. In the layer-by-layer growth, the depositing atoms or molecules are more strongly bonded to the substrate than each other and each layer is progressively less strongly bonded than the previous layer. This effect is continuous till the bulk bonding strength is reached. Layer-by-layer growth is hindered by elastic constraints at the solid-solid interfaces. A typical example of this is the epitaxial growth of semiconductors and oxide materials (Zexian 2011).

The SK model combines the feature of layer-by-layer and island growth. In this growth mode, after forming one or more monolayers in a layer-by-layer fashion, continued layer-by-layer growth becomes energetically unfavorable and islands begin to form. Due to the lattice mismatch between the substrate and the film, layer-by-layer growth cannot be accommodated for the higher thickness of the film and thus promotes the three-dimensional growth over the layer-by-layer growth. This type of growth is quite common and has been observed in several metals to metal and metal to semiconductor systems.

Though the film properties vary by the choice of deposition procedure, the following factors affect the growth, structure, and properties of a deposited film for any deposition method:

  i. Nature of the substrate (single crystal or polycrystalline)
 ii. Ratio of substrate temperature (Ts) and source temperature (Tm)
iii. Presence of vacuum and residual gases inside the chamber
 iv. Post annealing temperature and duration
  v. Impurities and the presence of defects on the substrate surface

Substrate temperature and film thickness are the two important parameters that influence the properties of the film. The adatom mobility on the substrate surface can be enhanced by increasing the substrate temperature (Schulz and Fritscher 1996). Higher mobility of the adatoms leads to a smaller nucleation barrier resulting in thin films with large crystallites sizes. Low mobility of adatom on the substrate surface at a lower substrate temperature leads to the formation of an amorphous film. Thus, a thin film can be deposited either in polycrystalline or amorphous nature by the choice of substrate temperature.

4.15 Importance and Applications of Coatings

Metal coatings provide significant contribution to different industrial sectors offering surface protection to many of the industrial components that we purchase. A metal coating is used to tailor the surface of the workpiece, which elevates the surface properties of the materials. Surface modification by composite materials achieved superior properties which cannot be attained if the workpiece is used alone without surface modifications. The surface coatings provide corrosion and wear resistance which furnishes load bearing property and better durability of the material. The metallic coatings such as chromium, cadmium, nickel, and copper are usually prepared by a wet chemical technique, which constitutes inherent impurities and growth controlling problems. Nanostructured metallic coatings such as chromium, nickel, cadmium, aluminum, and zinc are deposited by electroless plating, ion vapor deposition, electroplating, and chemical vapor deposition.

Electro/electroless plating is a kind of coating process, where the samples are generally dipped into the solutions containing metal salts, surfactants, and stabilizing agents, with subsequent removal from baths to obtain uniform and dense conditions. Similarly, coatings prepared by the sputtering technique exhibit better adhesion and a higher smooth surface of the film which is majorly used for optical industries. Several nitrides, carbides, and other ceramic coatings have been used in various industries including thin-film solar cells, wear-resistance coatings, and biosensor applications. In addition, copper, nickel, and silver coatings prepared by the DC/RF sputtering technique are increasingly used for electronic industries. Thermal barrier and diffusion barrier coatings (YSZ, CaSZ, and CeSZ) were developed by electron beam evaporation and plasma spray coatings.

The most important method of metallic coating for corrosion and wear protection is galvanizing that involves the application of metallic zinc to carbon steel for better corrosion and wear control. Hot-dip galvanizing is also the most common technique that constitutes dipping of steel samples into a bath containing molten zinc. Other modern techniques, such as plasma arc spraying are especially used for exotic refractory metals for versatile demanding applications. Hence, metallic coatings have been significantly used for various applications especially to improve the performance of the materials and durability.

References

Ana, Zuzuarregui, and Carmen Morant Minana Maria. 2016. *Research Perspectives on Functional Micro- and Nanoscale Coatings*, 34. Edited by Ana Zuzuarregui and Maria Carmen Morant-Minana. I. Advances in Chemical and Materials Engineering. IGI Global.

Ashassi-Sorkhabi, Habib, and Moosa Es'haghi. 2013. "Corrosion resistance enhancement of electroless Ni-P coating by incorporation of ultrasonically dispersed diamond nanoparticles." *Corrosion Science* 77 (December). Pergamon: 185–93.

Baumgärtner, M. E., Ch J. Raub, and D. R. Gabe. 1997. "Assessment of surface cleanliness for metal surfaces using an electrochemical technique." *Transactions of the Institute of Metal Finishing* 75 (3). Institute of Metal Finishing: 101–7.

Behera, Ajit, and S. C. Mishra. 2012. "Prediction and analysis of deposition efficiency of plasma spray coating using artificial intelligence method." *Open Journal of Composite Materials* 02 (02). Scientific Research Publishing: 54–60.

Christensen, Bjørn E. 1999. "Physical and chemical properties of extracellular polysaccharides associated with biofilms and related systems." In Jost Wingender, Thomas R. Neu, Hans-Curt Flemming (ed.), *Microbial Extracellular Polymeric Substances*, 143–54. Berlin Heidelberg: Springer.

Clerc, C., M. Datta, and D. Landolt. 1984. "On the theory of anodic levelling: Model experiments with triangular nickel profiles in chloride solution." *Electrochimica Acta* 29 (10). Pergamon: 1477–86.

Codony, Francesc, Jordi Morató, and Jordi Mas. 2005. "Role of discontinuous chlorination on microbial production by drinking water biofilms." *Water Research* 39 (9). Elsevier Ltd: 1896–906.

Cuthrell, Robert E. 1979. "The influence of hydrogen on the deformation and fracture of the near surface region of solids: Proposed origin of the rebinder-westwood effect." *Journal of Materials Science* 14 (3). Kluwer Academic Publishers: 612–18.

Duffy, John, Lester Pearson, and Milan Paunovic. 1983. "The effect of PH on electroless copper deposition." *Journal of The Electrochemical Society* 130 (4). The Electrochemical Society: 876–80.

Evans, U. R. 1969. "Mechanism of rusting." *Corrosion Science* 9 (11). Pergamon: 813–21.

Flemming, H. C. 2002. "Biofouling in water systems - cases, causes and countermeasures." *Applied Microbiology and Biotechnology* 59 (6): 629–640.

Fuchs-Godec, Regina, Miomir G. Pavlovic, and Milorad V. Tomic. 2015. "The inhibitive effect of vitamin-c on the corrosive performance of steel in HCl solutions - Part II." *International Journal of Electrochemical Science* 10 (12): 10502–12. www. electrochemsci.org.

Hagiwara, Tomoaki, Saki Hagihara, Akihiro Handa, Nobuyuki Sasagawa, Risa Kawashima, and Takaharu Sakiyama. 2015. "Pretreatment with citric acid or a mixture of nitric acid and citric acid to suppress egg white protein deposit formation on stainless steel surfaces and to ease its removal during cleaning." *Food Control* 53. Elsevier Ltd: 35–40.

Kamalan Kirubaharan, A.M., P Kuppusami, Akash Singh, T. Dharini, D. Ramachandran, and E Mohandas. 2015. "Structural and optical properties of electron beam evaporated yttria stabilized zirconia thin films." In *AIP Conference Proceedings*, 1665:080052. , UGC-DAE, CSIR, Indore.

Lehtola, Markku J., Michaela Laxander, Ilkka T. Miettinen, Arja Hirvonen, Terttu Vartiainen, and Pertti J. Martikainen. 2006. "The effects of changing water flow velocity on the formation of biofilms and water quality in pilot distribution system consisting of copper or polyethylene pipes." *Water Research* 40 (11). Elsevier Ltd: 2151–60.

Ludensky, Michael. 2003. "Control and monitoring of biofilms in industrial applications." *International Biodeterioration and Biodegradation*, 51:255–63 Elsevier Ltd.

Ma, Hongfang, Fang Tian, Dan Li, and Qiang Guo. 2009. "Study on the nanocomposite electroless coating of Ni-P/Ag." *Journal of Alloys and Compounds* 474 (1–2). Elsevier: 264–67.

Malecki, A., and A. Micek-Ilnicka. 2000. "Electroless nickel plating from acid bath." *Surface and Coatings Technology* 123 (1). Elsevier: 72–7.

Mallika, K, and R Komanduri. 1999. "Diamond coatings on cemented tungsten carbide tools by low-pressure microwave CVD." *Wear* 224 (2): 245–66.

Mandich, N. V. 2003. "Surface preparation of metals prior to plating: Part 1." *Metal Finishing*. Elsevier USA.

Mattox, D. M. 1996. "Surface effects on the growth, adhesion and properties of reactively deposited hard coatings." *Surface and Coatings Technology* 81 (1). Elsevier: 8–16.

Ning, Xian Jin, Jae Hoon Jang, and Hyung Jun Kim. 2007. "The effects of powder properties on in-flight particle velocity and deposition process during low pressure cold spray process." *Applied Surface Science* 253 (18): 7449–55.

Pauleau, Y. 2002. *Chemical Physics of Thin Film Deposition Processes for Micro- and Nano-Technologies*. Edited by Yves Pauleau. II. Dordrecht: Springer Netherlands.

PauliukaitĖ, R., G. Stalnionis, Z. Jusys, and A. Vaškelis. 2006. "Effect of Cu(II) ligands on electroless copper deposition rate in formaldehyde solutions: An EQCM study." *Journal of Applied Electrochemistry* 36 (11). Springer Netherlands: 1261–69.

Paunovic, Milan, and Mordechay Schlesinger. 2006. *Fundamentals of Electrochemical Deposition*, 2nd Edition. John Wiley & Sons, UK.

Petrov, I., P.B. Barna, L. Hultman, and J.E. Greene. 2003. "Microstructural evolution during film growth." *Journal of Vacuum Science & Technology A* 21 (5): S117.

Ranjan, Sarma. 2005. "Optoelectronic properties of thermally evaporated Zn Te thin films." Gauhati University. Sodhganga, http://hdl.handle.net/10603/68091

Reichelt, K. 1988. "Nucleation and growth of thin films." *Vacuum* 38 (12): 1083–99.

Schulz, U, and K Fritscher. 1996. "EB-PVD coatings- crystal habit and phase composition." *Surface & Coatings Technology* 82: 259–69.

Singh, J., and D. E. Wolfe. 2005. "Review nano and macro-structured component fabrication by Electron Beam-Physical Vapor Deposition (EB-PVD)." *Journal of Materials Science* 40 (1). Kluwer Academic Publishers: 1–26.

SreeHarsha, K. S. 2006. *Principles of Physical Vapor Deposition of Thin Films.* Elsevier. http://www.sciencedirect.com/science/book/9780080446998.

Subramanian, V., P. K. Bhattacharya, and A. A. Memon. 1995. "Chemical contamination of thin oxides and native silicon for use in modern device processing." *International Journal of Electronics* 78 (3). Taylor & Francis Group: 519–25.

van den Meerakker, J. E. A. M., and J. W.G. de Bakker. 1990. "On the mechanism of electroless plating. Part 3. Electroless copper alloys." *Journal of Applied Electrochemistry* 20 (1). Kluwer Academic Publishers: 85–90.

Zexian, Cao. 2011. *Thin Film Growth Physics, Materials Science and Applications.* Woodhead Publishing: Philadelphia.

5

Characterization of Surface Coatings

**K. Gobi Saravanan, A. M. Kamalan Kirubaharan,
and Vinita Vishwakarma**

CONTENTS

5.1 Introduction

There are several techniques available to characterize the bulk or surface-modified samples. Since surface coating plays a predominant role in the performance of the materials, this chapter emphasizes various tolls to characterize the deposited films. Providing specific answers to the queries such as composition, roughness, thickness, and other related parameters facilitates good quality coatings for desired applications. For instance, wear-resistant coatings require low hardness, which is close to the hardness of abrasive particles. In addition, the chromium content present in the chromium alloy is predominantly determined by the corrosion resistance of the alloy. Another prerequisite characterization of coatings is the quality of surface as the surface tailored materials may undergo defect as that of bulk materials.

Coating uniformity, film thickness, composition of the coatings, crystallinity, hardness, etc. play a major role in material durability. Cracks, voids, non-uniformity, pinholes, and residual stress in the coatings significantly damage material performance. Hence, some of the analytical techniques have been used predominantly to identify the crystallinity, morphology, and compositional changes of the coatings. The present chapter discusses different types of analytical techniques used to characterize the coatings.

5.2 Surface Analytical Techniques

5.2.1 X-Ray Diffraction

The crystallinity, lattice parameter of the phase, preferred orientation, and intrinsic stress of bulk as well as coatings were analyzed using X-ray diffraction (XRD) (Cullity 1956). XRD is a non-destructive technique and reveals major and minor phases present in the sample within the detection limit of ~5% of phase abundance (Conconi et al. 2014). The XRD pattern provides information on intensity versus diffraction angle. From the diffraction angle of a particular plane, the d-spacing and lattice parameters can be obtained. The strain and crystallite size are determined from the FWHM of a particular reflection after profile fitting. From the peak intensity, one can estimate the phase fractions present in the system (Prevéy 2000). XRD patterns of coatings were recorded using grazing incident geometry at an incident angle of 0.3°. The Gaussian fitting was used to extract FWHM values from the peak reflections. The 2θ positions were compared with JCPDS files of the respective materials published with star quality. From the diffraction angle and Miller indices, accurate lattice parameters were calculated using regression analysis combined with the nonlinear least-square fitting with the help of the 'Unit Cell program' (Holland and Redfern 1997). The crystallite size of a phase present in the coating was calculated from the full width at half maximum (FWHM) due to peak broadening. XRD can be effectively used to determine crystallite size in the range of 10–100 nm.

5.2.2 Residual Stress Measurement

The residual stress of thin films was analyzed by gracing incidence X-ray diffraction (GIXRD) measurement using a Bruker D8 Discover diffractometer with 2θ angle in the range of 20°–80° and Cu Kα radiation generated at 40 kV and 100 mA. The residual stress in the thin films was calculated using a modified sin2ψ technique after setting to three different angles of incidence (ω) (Ma, Huang, and Chen 2002). The residual stress gradient close to the film surface and film/substrate interface was determined using GIXRD at

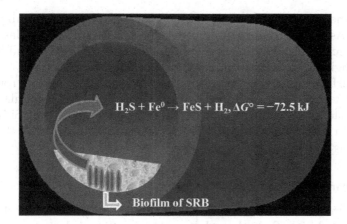

$$H_2S + Fe^0 \rightarrow FeS + H_2, \Delta G^\circ = -72.5 \text{ kJ}$$

Biofilm of SRB

FIGURE 5.1
Schematic representation of H_2S formation by sulfate-reducing bacteria (SRB).

different X-ray penetration depths by varying the angle of incidence. The angle of incidence was chosen based on the film thickness and penetration depth of the X-ray source into the material. The inter-planar spacing between crystal planes (d-value) with different Miller indices (hkl) was determined from respective peak positions obtained from GIXRD in the 2θ range between 20° and 90°. In the modified $\sin2\psi$ method (Figure 5.1), the incident beam angle to the specimen was fixed and the diffraction peaks of different (hkl) planes were collected in a single 2θ scan so that the sample tilting is not required like in the case of the conventional $\sin2\psi$ method (Ma, Huang, and Chen 2002).

5.2.3 Atomic Force Microscopy

Atomic force microscopy (AFM) is a non-destructive method used to analyze the surface topographic and roughness information about the sample. Unlike SEM, this technique does not require any special sample preparation step to make it electron conducting (Meyer 1992). In addition, an AFM study can be performed under ambient conditions. Three different modes are available in the AFM instrument: (i) contact mode, (ii) non-contact mode, and (iii) tapping mode (Jalili and Laxminarayana 2004). In the contact mode, the distance between the cantilever tip and the sample is about a few angstroms (Å) and the force between them is repulsive. In the non-contact mode, the force between the sample and cantilever tip is about some tens of angstroms and there is an attractive force between them. The intermittent contact between the tip and sample produces tapping mode images. The tip radius determines the resolution limit of the instrument. In this study, the tapping mode was used to analyze the coatings prepared by EBPVD. A Si_3N_4 cantilever tip was used to scan the coating surface by maintaining a constant force between the film surface and the cantilever tip. The laser beam could fall on the cantilever tip and the deflection was recorded by a photodiode while scanning on the sample. As the cantilever moved across the sample surface, the laser beam was deflected in the z-direction to maintain a constant force and the converted digital signals form an image.

The surface topography and root mean square (RMS) surface roughness of the coatings were examined with NTEGRA PRIMA (Modular Mode, Ireland) atomic force microscope (AFM) in tapping mode using Si_3N_4 cantilever with a tip radius of 50 nm. The typical scan range used for this study was 3 μm×3 μm.

5.2.4 Scanning Electron Microscopy

Electron interaction with samples has been extensively studied for characterizing the materials. Secondary electrons, backscattered electrons, and X-rays produced by the electron interactions carry information about the samples that are being investigated. A scanning electron microscope (SEM) uses electron beams to observe the sample morphology of a sample at higher magnification, higher resolution, and depth of focus (Tanaka and Maeda 2014). Herein, backscattered electrons, secondary electrons give information about the microstructure of the sample. X-rays are energetic photons produced from the sample and give information about the chemical identification, structure, and composition (Gotsch et al. 2016).

5.2.5 X-Ray Photoelectron Spectroscopy

X-ray photoelectron spectroscopy (XPS) is a key spectroscopic non-destructive technique for the surface characterization and analysis of the material (Sarma et al. 2013). This technique is also known as ESCA (Electron Spectroscopy for Chemical Analysis) and it offers a complete elemental analysis of the samples except for hydrogen and helium. XPS is surface sensitive (10–200 Å), and it also provides elemental and chemical bonding information more quantitatively. XPS requires high vacuum condition (10^{-7} to 10^{-8} Pa) and pre-sputtering has to be done before analysis for removing the surface contaminations. In XPS, the sample surface is irradiated by soft X-rays and the photoelectrons ejected from the sample are being analyzed. The energy of the emitted photoelectrons is essentially given by the difference between X-ray energy and binding energy of the electrons at their initial atomic energy levels, and this energy is characteristic of each element. By measuring the photoelectron energy with intensity, the elements and their abundances are recorded.

5.2.6 Optical Microscopy

The polished substrates and corrosion-tested coatings were used to examine under a monochromatic light in an upright microscope. Optical microscopy was employed on the Inconel-690 specimen after etching with an etchant (diluted HF solution) to examine the grain sizes and precipitates along the grain boundaries. It was also used to identify the type of corrosion on coated alloy specimens.

5.2.7 High-Resolution Transmission Electron Microscopy

Transmission electron microscopy (TEM) is the premier tool for understanding the internal microstructure of materials at the nanometer level. The resolution can be few tenths of nanometers, depending upon the imaging conditions, and can obtain electron diffraction patterns from the specific regions in the image as small as 1 nm. The instrumentation in TEM consists of an electron gun, a condenser lens, an objective lens, a magnification and projection lens, and a detector screen. The condenser lens is used to focus the electron beam produced and it controls the diameter of the electron beam. The objective lens is used to focus the image and helps to magnify the image. The images or the diffraction patterns are magnified and projected for viewing or recording through several lenses, which are the components of an imaging system.

There are two main types of imaging mode. The transmitted and diffracted beams can be recombined at the image plane, thus holding their phases and amplitude. This results in the phase-contrast image of the object. An amplitude contrast image can be obtained by eliminating the diffracted beams. This is achieved by placing suitable apertures below the back focal plane of the objective lens. This imaging mode is called the bright field image. The dark field image is obtained by using diffracted beams and by excluding the transmitted beam (Williams and Carter 2009).

High-resolution transmission electron microscopy (HRTEM) uses both transmitted and diffracted beams to create an interference image. It is a phase-contrast image and can be as small as the unit cell of the crystal. HRTEM has been extensively and successfully used for analyzing crystal structures and lattice imperfections in various kinds of advanced materials on an atomic resolution scale. It can be used for the characterization of point defects, stacking faults, dislocations, and precipitates at grain boundaries. However, the contrast in HRTEM images strongly depends on the sample thickness.

5.3 Characterization of Coatings

In order to elucidate the mechanism of high corrosion resistance performance of electroless plated Ni–P alloys, X-ray photoelectron spectroscopy (XPS) and electrochemical impedance analysis were studied by Elsener et al. (2008) to investigate specific details of optimized Ni–P coating.

Research analysis divulged that the enriched phosphorus layer formation at the alloy surface evolved a diffusion-controlled dissolution process.
Surface analysis of Ni-P alloy coated steel by XPS/XAES surface analysis
exhibited that phosphorus in the alloy surface had a similar chemical state
as elemental phosphorus. Furthermore, phosphorus in the bulk alloy was
negatively charged and form chemical interaction with nickel atoms, which
might facilitate the electronic structure, resulting in enhanced resistance to
dissolution in Ni–P alloys. At the same time, oxide type of passivity might
be ruled out owing to the fact that no nickel oxide was observed on polar-
ized Ni–P alloy.

Jin Xu et al. stated that SRB produced severe structural deformation and
corrosion products onto carbon steel Q235 (Xu et al. 2011). They immersed the
coupons for about 52 days in the presence and absence of SRB and observed
blow ball-like products seen onto a steel surface. They also observed pitting
holes with severe surface deformation of carbon steel Q235 containing SRB
rather than the absence of SRB. In addition, a sulfur component was observed
from SRB-treated carbon steel using X-ray photoelectron spectroscopy (XPS),
which confirmed the detrimental effects of SRB. Electrochemical impedance
spectroscopy (EIS) results of SRB on carbon steel evolved less charge transfer
resistance (Rct) rather with increasing time duration (from day 1 to 52 days)
due to that pits formation on the surface influenced less Rct taking place
(Xu et al. 2011).

The biofilm matrices and chemical composition vary significantly depending on the type of microbial cells and their metabolic activity, physicochemical conditions (shear force, temperature, pH, etc.), and different stages of
biofilm development ((i) irreversible attachment, (ii) microcolony formation,
(iii) development, (iv) maturation of biofilm, and (v) detachment of bacterial
cells). In addition, acquiring information from EPSs will give new insight
to develop novel antibiotics, find out new diagnostic tools, detergent agents,
biocide agents, and optimization of antifouling strategies. Analytical methods are an effective tool to identify organic compounds such as nucleic
acids, proteins, and hydrocarbons, to elucidate their structures. Knowledge
of microscopy is one of the pivotal techniques used to analyze biofilms. A
confocal laser scanning microscope (CLSM) is used to monitor different
steps involved in bacterial adherence and its dynamics during the accumulation (Kristensen et al. 2006). During initial bacterial attachment and subsequent biofilm formation, the organic and inorganic substances are secreted
by bacterial cells which are commonly analyzed and visualized by microscopic, spectroscopic, and molecular techniques, gram staining methods,
etc. (Lewandowski and Beyenal 2007). Some conventional chemical and biochemical methods such as genetic assays, nucleic acid assay, proteomics, and
fatty acid profiling are commonly used. However, the conventional methods
are labor intensive and require state-of-the-art instrumental facilities with
highly trained personnel.

FISH is a recognized tool for specific and sensitive identification of organisms within microbial complex communities. The main objective of the FISH analysis is to develop a probe for 16S rRNA target-specific sequences and this could be combined with confocal laser scanning microscopy (CLSM) for the analysis of gram-positive and gram-negative organisms with spatial distribution. After hybridization, the labeled bacterial cells in the microbiota can be determined by ribosomal RNA-targeted nucleic acid probes. Among all the microscopy techniques, CLSM is effectively used due to its non-destructive and three-dimensional visualization of bacterial cells and reconstruction of biofilms without structural distortion by a computational approach. However, the most significant experimental problem needs to be noticed that the cell walls of gram-positive bacteria should be penetrable to the probes without loss of signals from gram-negative organisms, which are labeled with fluorescent nucleotide probes. Moreover, continuous hybridization should not harm biofilm structures. However, the experimental parameters are quite sensitive and complicated to analyze total EPS staining in CLSM.

For better visualization of biofilm, SEM offers higher resolution, which facilitates distinct bacterial morphology on a nanometer scale. EPS and inorganic precipitation of biofilm can be clearly viewed. For SEM observation, the sample preparation for bacterial biofilm involves specific fixation procedures needed to retain the appropriate bacterial structure without any damage during image processing (Alhede et al. 2012). Briefly, the biofilm samples were gently washed with saline followed by a glutaraldehyde solution for about 3 hours. After incubation, biofilm samples are again gently washed with saline solution and then treated with different concentrations of alcohol ranging from 20, 40, 60, 80, to 100. Finally, the samples were cold dried and sputtered by gold alloy prior to SEM observation (Singh and Wolfe 2005). However, environmental SEM or cryo-SEM are used in order to prevent the collapse in bacterial morphology as well as EPS matrix especially for unfixed samples (Vertes, Hitchins, and Phillips 2012). Another characterization method is optical coherence tomography, which comprises low-level spatial resolution than SEM. Though it expresses low resolution, *in situ* three- and two-dimensional biofilm structures can be viewed without the destruction of samples.

In addition, atomic force microscopy (AFM) is an effective technique by which surface topography along with physical and mechanical (force curve measurement) properties of microbial systems can be analyzed (Wright et al. 2010). Depth-resolved analysis or photoacoustic technique is an important method to detect biofilm density or thickness (mature biofilm) in which more than 1 cm thickness of the biofilm can be measured. Nuclear magnetic resonance spectroscopy (NMR) is also used for characterizing biofilm and its structural analysis. Furthermore, different types of spectroscopic methods such as matrix-assisted laser desorption ionization MS (mass spectroscopy), secondary ion MS, and desorption electrospray ionization MS are available for characterization, imaging, and identification of biofilms. However, biofilm analysis by MS is time consuming, expensive, and requires complicated sample preparation methods. Additionally, quantification of imaged molecular species of biofilms is also quite tedious in this MS technique (Vertes, Hitchins, and Phillips 2012).

5.4 FTIR Spectroscopy

An early microbial investigation was used to study by Raman microspectroscopy, which mainly focused on the diffusion process. Suci analyzed the temporal and spatial structural distribution of chlorhexidine in the presence of *Candida albicans* biofilm by Raman and attenuated total reflection FTIR spectroscopy (Suci, Geesey, and Tyler 2001). In *Streptococcus mutans* biofilm, polyethylene glycol diffusion was studied by Marcotte, Barbeau, and Lafleur (2004). The biofilm distribution studied at $2,900\,cm^{-1}$ wavelength belongs to CH stretching. From the observation, they found that heterogeneous *S. mutans* biofilm distribution was found for polyethylene glycol, which will vary depending on the biomass content.

To characterize the biofilm, Raman spectroscopy plays a vital role to obtain several information such as (i) biofilm formation and its development, (ii) variations in chemical composition due to stress and metabolic activity in the biofilm, and (iii) chemical heterogeneity in the biofilm with respect to physiologic state and deterioration of biofilm due to antibacterial agents.

The study of microbial growth and its heterogeneity was analyzed in detail (Choo-Smith et al. 2001). The results revealed that biofilm surfaces exhibited a high level of glycogen; at the same time, a high level of RNA was present in the deeper layer exhibiting higher metabolic activity in an inner layer, which plays an important role in the antibiotic resistance mechanism. Sandt et al. (2007) intensively used Raman spectroscopy to study chemical heterogeneity, structure, and composition and developed a fully hydrated biofilm on glass in a flow cell. Later the same group found that Raman spectroscopy is also used to obtain information about the microbial species or strain-related variation of biomass density and water content in *Pseudomonas aeruginosa* and *Pseudoalteromonas* sp. biofilms. They analyzed the O–H stretching band at $3,405\,cm^{-1}$ (water) and C–H stretching bands (biomass) at $2,950\,cm^{-1}$ to find the biomass water content variations in biofilms. Huang et al. examined the effectiveness of Raman spectroscopy in analyzing the chemical composition difference between *Pseudomonas fluorescence* cells that are cultivated under different stress conditions with metabolic histories (Huang et al. 2007). The chemical composition of whole cells showed differences in different carbon source media rather than in the starvation period.

The important and considerable drawback of Raman spectroscopy is the low quantum efficiency, which leads to less Raman effect (10^{-8} to 10^{-6}) resulting in limited sensitivity. Hence, very few biofilm measurement spots could be analyzed. To improve the quantum efficiency, powerful lasers and detectors will be needed. Hence, techniques can be increasingly needed to enhance the Raman Effect to a better understanding of the chemical composition and heterogeneity of the samples. In this, gold and silver nanostructured materials are used to enhance the Raman Effect or by resonance Raman effect. The surface-enhanced Raman scattering (SERS) is the primary enhancement process in which chemical and electromagnetic enhancement was performed to obtain higher sensitivity. The SERS occurs when the incident photon energy is equal to or close to the electronic transition energy of the molecule. Therefore, when a biological sample comprises the Raman active resonances (chromophores), a rapid Raman scattering is possible, if the appropriate wavelength is fixed.

A biofilm consists of several types of microorganisms, from which, it is very tedious to acquire chemical composition and heterogeneity of species. To get information from single bacterial cells, the researchers found that Raman spectroscopy is an effective tool to acquire more information about biofilms. Schuster et al. reported for the first time about the biofilm by Raman microspectroscopy in 2000 (Schuster, Urlaub, and Gapes 2000). They investigated the chemical composition and heterogeneity of *Clostridium acetobutylicum*. Maquelin et al. demonstrated the Raman spectroscopy for species and even strain-specific discrimination of different microbial pathogens (Maquelin et al. 2003). There is some technology such that Raman signals can be amplified by which acquisition time was reduced. Resonance Raman microspectroscopy is increasingly used to identify the presence of cytochrome c in electroactive bacterial cells. Electroactive bacteria play an important role in the production of electron transfer or electricity processes.

Raman signals of analytes can be improved if they locate remarkably close to the desired wavelength or attached to nanostructured metallic materials (gold or silver). During Raman analysis, the SERS spectra of whole bacteria will exhibit from the surface of the cell membrane to analyze their metabolic activity or molecular species identification (Kahraman et al. 2008). It is known that emission SERS spectra can be sensitive to a molecule of interest in a cell membrane. For instance, the SERS signals of flavin adenine dinucleotide (FAD) are excited at 514.5 nm (Zeiri et al. 2002). Kubryk et al. studied the origin of the strong band originated from bacteria by SERS using stable isotope labeling and found around 730 cm^{-1} (Kubryk, Niessner, and Ivleva 2016).

Premasiri et al. (2016) wanted to study the biochemical origin of bacteria by SERS. A 785 nm laser was used with gold nanoparticle-covered SiO$_2$ as the substrate material. They mainly focused on 'purine', which is a dynamic molecular species for biochemical origin in surface-enhanced Raman spectra.

From the observation, they evidently proved the active metabolic purine pathway reaction (Premasiri et al. 2016). The metallic colloidal nanoparticles are used to analyze the interior bacterial cells using enzymatic reduction of metallic salts (gold/silver). The SERS method remains challenging due to non-reproducibility spectra from bacterial cells. In order to improve the accuracy of SERS signals, Kahraman et al. (2008) conducted a research with low-magnification lenses for dried cells or cell suspensions to analyze a group of cells at once or individual cells to acquire good enhancement in reproducibility. The different signals from SERS with low magnification in the presence of many bacteria acquire higher specificity. In addition, optimized SERS substrate can effectively improve reproducibility by a single bacterial cell using high magnification lenses.

Ramya et al. (2010) observed the chemical composition in biofilms and EPS of *P. aeruginosa* and algae grown in titanium metal surfaces using Raman spectroscopy. In addition, they used SERS experiments in the presence of mono (silver/copper) and bimetallic colloids (silver and copper) with the laser wavelength of 633 nm. From the results, they identified that the Raman spectrum with SERS showed better accuracy for identification, quantification, and differentiation of bacteria and algae. Chao and Zhang used the SERS technique to identify the variations in chemical composition during different growth stages of *E.coli* (gram (-)), *Bacillus subtilis*, and *Pseudomonas putida* (gram (+)) organisms with silver colloids. A laser excitation wavelength of 633 nm was fixed during analysis and the results exhibited that nucleic acids, lipids, and protein contents have increased significantly during cultivation. The obtained SERS spectra are sorted into different macromolecular classes (Chao and Zhang 2012).

5.5 Biofilm Analysis in Membrane Water Filters

Membranes play a vital role in water purification/filtration industries to improve the quality of water. Since the microbial invasion is a major challenge in the membrane system, there is a need to analyze the chemical composition and other heterogeneity of the microbial invasion, which will pave the way for corresponding membrane cleaning processes to inhibit biofilm development. Gold nanoparticles obtained by reduction of gold salt (Chloroauric acid) were effectively used to monitor the development of biofilm in dual bacterial species (*Brevundimonas diminuta* and *Staphylococcus aureus*) by Raman spectra coupled with a Surface-Enhanced Raman Scattering (SERS) system. The changes in the biofilm and the quantity of biomass were demonstrated with the laser excitation wavelength of 633 nm (Chen, Cui, and Zhang 2015). The same group also studied about the chemical composition variation during biofilm development as well as during cleaning by layer-by-layer SERS analysis (Cui et al. 2015). From the observation, they stated that the technique brings new insight to identify the surface of the biofilm, which is directly exposed to cleaning reagents by increasing or decreasing with Raman bands. Furthermore, real-time online detection of biofouling in drinking water membranes by the SERS method was demonstrated. For this, they fabricated the SERS sensor in the presence of gold nanoparticles embedded onto the membrane filters to detect a very low concentration of surface biofoulants in a real-time environment, and soon it was immersed in ultra-high pure water systems (Kögler et al. 2016). The developed gold nanoparticles-based real-time SERS sensing for online monitoring of biofoulant detection is a suitable approach in the present scenario especially for high water flux with pressure. However, fluorescent-based SERS sensing in real-time applications still needs to be investigated.

Hence, the SERS spectroscopy is an established and powerful method to overcome the sensitivity limitations and chemical composition, which favors obtaining target specific and more detailed chemical composition on biofilm rather than the classic Raman spectroscopy.

5.6 Identification of Chemical Composition in Biofilm Formation

The intrinsic chemical composition of biofilm can be analyzed by two different vibrational techniques namely Raman spectroscopy and Fourier transform IR spectroscopy (FTIR) (Sharma and Prakash 2014). In Raman spectroscopy, vibrations change the polarizability of the samples which exhibit visible regions in the spectrum. In Raman spectroscopy, the formation of photons during scattering transfers some energy that corresponds to the specific molecule vibrations. In IR spectroscopy, vibrations make changes in the dipole moment in the material or molecules (Noothalapati Venkata, Nomura, and Shigeto 2011). During analysis, the molecular vibrations and excitations are involved during light scattering or by inelastic scattering of light. Two different vibrations, namely, Stokes and anti-Stokes vibrations, are majorly involved in the scattering. When the scattered light frequency is lower than the frequency of incident light, then Stokes scattering will be formed, whereas, when the frequency of scattering light is higher than the frequency of incident light, anti-Stokes scattering is formed. The physical difference between both methods leads to different intensities in the same vibrational band. For instance, OH stretching vibrations are more pronounced in IR spectra, whereas the same analysis is weakly visible in Raman spectra.

5.7 Impact of SRB on Pipelines

Microbiologically induced corrosion (MIC) is one of the important crises in the gas and oil industry. Microorganisms generally need four elements such as water, carbon, electron donor, and acceptor (Littile and Lee 2007). Hydrocarbons are the backbone of oil transporting pipelines, which is an abundant carbon source for microorganisms. To minimize carbon steel corrosion in pipelines, effective methods have been carried out using cathodic protection and surface engineering. MIC is one among the factors influencing corrosion caused by bacteria, fungi, microalgae, and archaea. Microbes can accelerate anodic/cathodic corrosion kinetic reactions by 'catalytic' entities. In particular, microbes pave the way to cause de-alloying, pitting, and stress corrosion cracking (Littile and Lee 2007; Lekbach et al. 2018; Li et al. 2000). Industrial experts state that billions of dollars have been lost due to oil and gas pipeline rupturing due to localized corrosion resulting in oil spillage leading to a large-scale environmental and ecological problem (Oyediran and Abraham 2005).

Pipelines and fittings are considered strategic components in petrochemical plants especially in oil/gas pipelines. Elbows are a major component in the pipeline industries, where severe deformation will usually occur owing to chemical and biological interaction. Among the various biological interactions, sulfate-reducing bacterial (SRB) interaction plays a vital role in pipeline industries that adhered onto the pipeline surface and grow in an anaerobic environment. Most of the pipelines are made of carbon steel that needs to be buried into the soil, which causes a deterioration process due to its poor corrosion resistance nature. The biological and chemical attack on the carbon steel surface is obtained from the external and internal environment of the steel. For the external environment, pH of the soil and its resistivity, soil chemistry, temperature, and microbiota play an important role in the corrosion process, which leads to environmental contaminants imposing health hazards to humans (Shi et al. 2018; Usher et al. 2014; Jia et al. 2019; Kiani Khouzani, Bahrami, and Eslami 2018).

The smooth and uninterrupted flow of oil and gas pipelines is the major challenge to pipeline engineers, field operators, and designers as the pipelines and fitting components undergo material deterioration due to various types of corrosion. The corrosion kinetics will be elevated in the oil/sag pipeline due to the presence of anode, cathode, and electrolyte. An aqueous environment acts as an electrolyte that transfers the electrons amidst anode and cathode and the structural component (pipeline) as anode where corroding of metal will be taking place. Finally, the cathodic region works as an electrical conductor in a cell that is not involved in the corrosion process (Corbin and Willson 2007).

Corrosion causes severe deformities in various fields like pressure basins, boiler tanks, turbine blades and motors, aggressive/harmful chemical containers, bridges, aerospace components, and automobile industries. Several researchers across the globe conducted surveys about the economic loss of a particular country due to corrosion. Jayaherdashti conducted a survey about the maintenance cost of corrosion-related issues in the oil, gas, and automobile industries and stated that most of the countries contribute 1%–5% of their gross national product (GNP) for corrosion-related problems every year (Javaherdashti 2008). The National Association of Corrosion Engineers (NACE) performed a study about the significance of corrosion and its preventive measures worldwide in 2013 and analyzed that the approximate estimation of corrosion cost is 3.4% of GDP (Koch, Varney, Thompson, Moghissi 2012). Lim explained the impact of corrosion in the Gulf Corporation Council (GCC) and suggested that the kingdom of Saudi Arabia (KSA) has undergone severe corrosion-related problems and spent the highest expenditure to tackle corrosion (Lim 2017). Besides, UK spent about £13.65 in 1969 (Okamoto, Schlesinger, and Mueller 1992). NACE and USFHWA conducted a 2-year study in 2002 and illustrated that the country spent about US\$ 276 B (3.1% of the US\$), which was nearly half of the amount allotted for corrosion-related maintenance issues, particularly the selection of high corrosion resistance alloys, mechanically and thermally stable polymers, corrosion inhibitors, surface modification of pipelines, and cathodic protectors (Tiu and Advincula 2015). Among these, oil and gas production energy industries contribute major cost for corrosion (Finšgar and Jackson 2014). The overall survey divulges that the oil and gas industry contributes approximately US\$ 1.4 B; petrochemical and chemical manufacturing industries spent about US\$ 1.7 B and US\$ 1.7 B for petroleum refinery industries (Finšgar and Jackson 2014). In general, crude oil and natural gas transporting pipelines consist of impurities such as water, carbon dioxide (CO_2), and other impurities formed by varying conditions of well owing to changes in temperature and pressure, compositional changes, and souring of well over the period of time.

It is known that approximately 20%–40% of localized pipeline corrosion problems were obtained by MIC. This ratio is more pronounced in the inner pipeline by 70%–95% caused by MIC (Javaherdashti 2008). SRB can be found in all forms of oil and gas-related products ranging from well to pipeline distribution, refineries, transport, and treatment facilities. SRB reduces sulfate ions to sulfide (H_2S) by a terminal electron acceptor process through an eight-electron transfer reaction (Figure 5.1).

The reduced sulfide product will act as a powerful anodic and cathodic corrosive reactant with metal surfaces (Hamilton 1985).

The bacteria present in the outer as well as the inner surface of carbon steel pipeline (oil, diesel, and petroleum products) can adhere to the surface by a planktonic mechanism where the initial biofilm formation was obtained. Once the bacteria are firmly attached to the surface, several stages will occur such as rapid growth, reproduction, nutrient consumption followed by secretion of extracellular polysaccharides, and other byproducts formation influences localized corrosion and it is collectively known as 'biofilm'.

It is well known that microbes use chemicals and oxidize them in such a way that ferrous ions present in the carbon steel will be oxidized to ferric oxide, thereby influencing corrosion in the absence of nutrients. The progression of microbial corrosion on steel is usually classified into aerobic and anaerobic microorganisms. Anaerobic microbes elevate a comparatively higher corrosion rate rather than aerobic ones. In the meantime, anaerobic organisms can control the control kinetics of corrosion in the presence of oxygen. The various microorganisms, namely, iron-reducing, iron and manganese-oxidizing, acid-producing, and sulfur-reducing bacteria, play pivotal roles in causing a detrimental effect to carbon steel. SRB are usually found in water, soil deposits, chemicals, hydrocarbons, etc.

Among these, SRB available in both external and internal surfaces of carbon steel pipeline cause major surface deformation in buried pipelines. In the internal environment, SRB tends to adhere inside the pipeline where stagnant liquid and low oxygen content influence SRB to cause a ten times faster corrosion rate than the absence of microorganisms. It is well known that SRB could directly intake Fe when there is no availability of carbon source in the surrounding environment. During the growth phase of SRB, it can utilize elemental irons as an energy source (e$^-$ donor). Furthermore, SRB consumes sulfate as a terminal e$^-$ acceptor at the cathodic part (Figure 5.2).

Furthermore, research has been focusing on corrosion monitoring and control of MIC in amine industry plants. The sweetening process in an amine plant evolves excess of hydrogen sulfide and carbon dioxide from natural or soar gas. Though numerous microorganisms play an important role in the

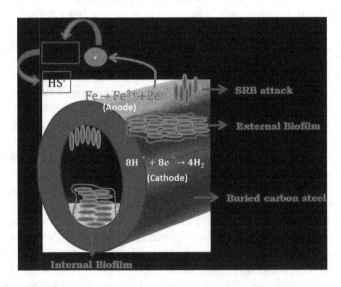

FIGURE 5.2
Schematic diagram of microbiologically induced corrosion (MIC) by sulfate-reducing bacteria (SRB).

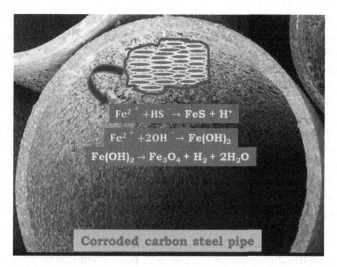

FIGURE 5.3
Schematic diagram of the mechanism of SRB in corroded carbon steel pipelines.

corrosion of buried pipelines, SRB take predominant role in contributing to MIC (Figure 5.3).

SRB cause localized pitting corrosion on buried carbon pipes generated from the outer surface. Khouzani et al. (2019) studied MIC by SRB in the elbow of the amine pipeline. The failure was observed on the amine pipeline due to the pitting corrosion caused by SRB. The observed pits were analyzed with three different morphologies, namely, rod-like, flower shape, and a colony of sphere-like morphologies. They also studied about the inner surface of the amine pipeline and found severe signs of localized corrosion. In addition, the inner surface was identified with a different morphology of porous scale-like structure, α-FeOOH particle sand honeycomb-like structures indicated that the inner surface was formed with iron oxide, hydroxides, and other oxides. However, no signs of chloride and sulfur were observed in the inner surface. Finally, they observed that the deleterious effect of SRB was more pronounced in the outer surface of the amine pipeline and corrosion products were observed in the inner surface due to the low level of oxygen and liquid present in the pipeline (Figure 5.3).

5.8 Biofouling Treatment and Preventive Measures

Chlorine is the most common biocide used in cooling towers. In addition to chlorine, copper has been used in certain ways for controlling bacterial growth. In addition, some antimicrobial agents such as copper (2–4 mg/l), silver (0.2–0.4 mg/l), and zinc nanoparticles (NPs) are also effective in removing and washing bacteria in cooling towers. In heat exchanger water circuits, 1 mg of free chlorine ions are required to arrest planktonic bacterial growth. At the same time, chlorine ions need four-fold higher concentrations to arrest the growth of sessile cells (Kim et al. 2002). Over the period of time, the bioaccumulation or biofouling effect on the surface of the pipelines facilitates corrosion and reduction in their working efficiency (Rondum et al. 2012).

The progress of corrosion influenced by microbes depends on the conditioning of the film. During the microbial invasion, the slimy layer of biofilms consists of not only organic compounds but also some inorganic compounds available in the matrix. The biofilm attracts other bacteria to multiply and grow, producing extracellular polysaccharides (ECM) resulting in the biofilm formation. After biofilm formation, the ion and oxygen concentrations inside the biofilm differ from those of the outer surface, which triggers corrosion on the pipelines.

To increase the durability of oil, gas, and petrochemical industries, (i) construction of equipment/pipelines with proper construction materials for specific applications, (ii) appropriate modification in corrosive media, and (iii) developing barrier amidst metal and medium by surface modification onto a metal surface play crucial roles for better success.

References

Alhede, Morten, Klaus Qvortrup, Ramon Liebrechts, Niels Høiby, Michael Givskov, and Thomas Bjarnsholt. 2012. "Combination of microscopic techniques reveals a comprehensive visual impression of biofilm structure and composition." *FEMS Immunology and Medical Microbiology* 65 (2). 335–42.

Chao, Yuanqing, and Tong Zhang. 2012. "Surface-Enhanced Raman Scattering (SERS) revealing chemical variation during biofilm formation: From initial attachment to mature biofilm." *Analytical and Bioanalytical Chemistry* 404 (5). 1465–75.

Chen, Pengyu, Li Cui, and Kaisong Zhang. 2015. "Surface-enhanced Raman spectroscopy monitoring the development of dual-species biofouling on membrane surfaces." *Journal of Membrane Science* 473 (January). Elsevier B.V.: 36–44.

Choo-Smith, L. P., K. Maquelin, T. Van Vreeswijk, H. A. Bruining, G. J. Puppels, N. A. Ngo Thi, C. Kirschner, et al. 2001. "Investigating microbial (micro) colony heterogeneity by vibrational spectroscopy." *Applied and Environmental Microbiology* 67 (4). Appl Environ Microbiol: 1461–69.

Conconi, M. S., M. R. Gauna, M. F. Serra, G. Suarez, E. F. Aglietti, and N. M. Rendtorff. 2014. "Quantitative firing transformations of a triaxial ceramic by X-ray diffraction methods." *Cerâmica* 60 (356). Associação Brasileira de Cerâmica: 524–31.

Corbin, Darryl, and Elfriede Willson. 2007. "New technology for real-time corrosion detection." In *Tri-Service Corrosion Conference*. December 3–6, United states of America, 2007.

Cui, Li, Pengyu Chen, Bifeng Zhang, Dayi Zhang, Junyi Li, Francis L. Martin, and Kaisong Zhang. 2015. "Interrogating chemical variation via layer-by-layer SERS during biofouling and cleaning of nanofiltration membranes with further investigations into cleaning efficiency." *Water Research* 87 (December). Elsevier Ltd: 282–91.

Cullity, B. D. 1956. *Elements Of X-Ray Diffraction.* Edited by B. D. Cullity. Addison-Wesley. Metallurgy Series, Reading, Massachusetts, United States.

Elsener, Bernhard, Maura Crobu, Mariano Andrea Scorciapino, and Antonella Rossi. 2008. "Electroless deposited Ni-P alloys: Corrosion resistance mechanism." *Journal of Applied Electrochemistry* 38 (7). Springer: 1053–60.

Finšgar, Matjaž, and Jennifer Jackson. 2014. "Application of corrosion inhibitors for steels in acidic media for the oil and gas industry: A review." *Corrosion Science* 86: 17–41.

Gotsch, Thomas, Wolfgang Wallisch, Michael Stoger-Pollach, Bernhard Klotzer, and Simon Penner. 2016. "From zirconia to yttria: sampling the ysz phase diagram using sputter-deposited thin films." *AIP Advances* 6 (2): 025119.

Hamilton, W A. 1985. "Sulphate-reducing bacteria and anaerobic corrosion." *Annual Review of Microbiology* 39 (1):195–217.

Holland, T. J. B., and S. A. T. Redfern. 1997. "Unit cell refinement from powder diffraction data: The use of regression diagnostics." *Mineralogical Magazine* 61: 65–77.

Huang, Wei E., Mark J. Bailey, Ian P. Thompson, Andrew S. Whiteley, and Andrew J. Spiers. 2007. "Single-cell Raman spectral profiles of pseudomonas fluorescens SBW25 reflects in vitro and in planta metabolic history." *Microbial Ecology,* 53:414–25.

Jalili, Nader, and Karthik Laxminarayana. 2004. "A review of atomic force microscopy imaging systems: Application to molecular metrology and biological sciences." *Mechatronics* 14 (8). Pergamon: 907–45.

Javaherdashti, Reza. 2008. "Microbiologically Influenced Corrosion (MIC)." In Derby, Brain (Ed.), *Microbiologically Influenced Corrosion,* 29–71. Verlag-London: Springer.

Jia, Ru, Tuba Unsal, Dake Xu, Yassir Lekbach, and Tingyue Gu. 2019. "Microbiologically influenced corrosion and current mitigation strategies: A state of the art review." *International Biodeterioration and Biodegradation.* Elsevier Ltd.

Kahraman, Mehmet, M. Müge Yazici, Fikrettin Şahin, and Mustafa Çulha. 2008. "Convective assembly of bacteria for surface-enhanced raman scattering." *Langmuir* 24 (3). American Chemical Society: 894–901.

Khouzani, Mahdi Kiani, Abbas Bahrami, Afrouzossadat Hosseini-Abari, Meysam Khandouzi, and Peyman Taheri. 2019. "Microbiologically influenced corrosion of a pipeline in a petrochemical plant." *Metals* 9 (4): 1–14.

Kiani Khouzani, M., A. Bahrami, and A. Eslami. 2018. "Metallurgical aspects of failure in a broken femoral HIP prosthesis." *Engineering Failure Analysis* 90 (August). Elsevier Ltd: 168–78.

Kim, B. R., J. E. Anderson, S. A. Mueller, W. A. Gaines, and A. M. Kendall. 2002. "Literature review - efficacy of various disinfectants against legionella in water systems." *Water Research.* Elsevier Ltd.

Koch GH, Varney J, Thompson N, Moghissi O, Gould M. 2012. *International Measures of Prevention, Application, and Economics of Corrosion Technologies Study.* Houston, TX: NACE International.

Kögler, Martin, Bifeng Zhang, Li Cui, Yunjie Shi, Marjo Yliperttula, Timo Laaksonen, Tapani Viitala, and Kaisong Zhang. 2016. "Real-time raman based approach for identification of biofouling." *Sensors and Actuators, B: Chemical* 230 (July). Elsevier B.V.: 411–21.

Kristensen, David M., Thomas R. Jørgensen, Rasmus K. Larsen, Mads C. Forchhammer, and Kirsten S. Christoffersen. 2006. "Inter-annual growth of arctic charr (Salvelinus Alpinus, L.) in relation to climate variation." *BMC Ecology* 6 (August). BioMed Central Ltd.: 1–8.

Kubryk, Patrick, Reinhard Niessner, and Natalia P. Ivleva. 2016. "The origin of the band at around $730\,cm^{-1}$ in the SERS spectra of bacteria: A stable isotope approach." *Analyst* 141 (10). Royal Society of Chemistry: 2874–78.

Lekbach, Yassir, Dake Xu, Soumya El Abed, Yuqiao Dong, Dan Liu, M. Saleem Khan, Saad Ibnsouda Koraichi, and Ke Yang. 2018. "Mitigation of microbiologically influenced corrosion of 304L stainless steel in the presence of pseudomonas aeruginosa by cistus ladanifer leaves extract." *International Biodeterioration and Biodegradation* 133 (September). Elsevier Ltd: 159–69.

Lewandowski, Zbigniew, and Haluk Beyenal. 2007. *Fundamentals of Biofilm Research*. *Fundamentals of Biofilm Research*. Boca Raton, FL: CRC Press.

Li, Seonyeob, Younggeun Kim, Kyung Soo Jeon, and Youngtai Kho. 2000. "Microbiologically influenced corrosion of underground pipelines under the disbonded coatings." *Metals and Materials International* 6 (3). Korean Institute of Metals and Materials: 281–86.

Lim, HL. 2017. *Materials Science and Engineering: Concepts, Methodologies, Tools, and Applications*. Vol. 1–3. IGI Global.

Littile, Brenda J., and Jason S. Lee. 2007. *Microbiologically Influenced Corrosion*. Wiley & Sons.

Ma, C. H., J. H. Huang, and Haydn Chen. 2002. "Residual stress measurement in textured thin film by grazing-incidence X-ray diffraction." *Thin Solid Films* 418 (2): 73–8.

Maquelin, K., C. Kirschner, L. P. Choo-Smith, N. A. Ngo-Thi, T. Van Vreeswijk, M. Stämmler, H. P. Endtz, H. A. Bruining, D. Naumann, and G. J. Puppels. 2003. "Prospective study of the performance of vibrational spectroscopies for rapid identification of bacterial and fungal pathogens recovered from blood cultures." *Journal of Clinical Microbiology* 41 (1). J Clin Microbiol: 324–29.

Marcotte, Lucie, Jean Barbeau, and Michel Lafleur. 2004. "Characterization of the diffusion of polyethylene glycol in streptococcus mutans biofilms by Raman microspectroscopy." *Applied Spectroscopy* 58 (11). Appl Spectrosc: 1295–301.

Noothalapati Venkata, Hemanth Nag, Nobuhiko Nomura, and Shinsuke Shigeto. 2011. "Leucine pools in *Escherichia Coli* biofilm discovered by Raman imaging." *Journal of Raman Spectroscopy* 42 (11). John Wiley & Sons, Ltd: 1913–15.

Okamoto, Hiroaki, Mark E. Schlesinger, and Erik M. Mueller. 1992. *ASM Handbook Volume 3: Alloy Phase Diagrams - ASM International*. Ohio: ASM International.

Oyediran, Benjamin, and A. Abraham. 2005. "Mathematical modeling : An application to corrossion in a petroleum industry." *NMC Proceeding Workshop on Environment*, National Mathematical Centre, Abuja, Nigeria, 48–56.

Premasiri, W. Ranjith, Jean C. Lee, Alexis Sauer-Budge, Roger Théberge, Catherine E. Costello, and Lawrence D. Ziegler. 2016. "The biochemical origins of the surface-enhanced raman spectra of bacteria: A metabolomics profiling by SERS." *Analytical and Bioanalytical Chemistry* 408 (17). Springer Verlag: 4631–47.

Prevéy, Paul S. 2000. "X-ray diffraction characterization of crystallinity and phase composition in plasma-sprayed hydroxyapatite coatings." *Journal of Thermal Spray Technology* 9 (3). Springer-Verlag: 369–76.

Ramya, S., R. P. George, R. V.Subba Rao, and R. K. Dayal. 2010. "Detection of algae and bacterial biofilms formed on titanium surfaces using micro-raman analysis." *Applied Surface Science* 256 (16). Elsevier B.V.: 5108–15.

Rondum, K. D., A. C. Cabell, E. S. Beardwood, and J. Krupp. 2012. "Performance-based cooling water treatment." *Power* 156 (7). McGraw Hill Inc.: 46–6. https://www.cheric.org/research/tech/periodicals/view.php?seq=1030249.

Sandt, C., T. Smith-Palmer, J. Pink, L. Brennan, and D. Pink. 2007. "Confocal Raman microspectroscopy as a tool for studying the chemical heterogeneities of biofilms in situ." *Journal of Applied Microbiology* 103 (5):1808–20.

Sarma, D. D., Pralay K. Santra, Sumanta Mukherjee, and Angshuman Nag. 2013. "X-ray photoelectron spectroscopy: A unique tool to determine the internal heterostructure of nanoparticles." *Chemistry of Materials* 25 (8). American Chemical Society: 1222–32.

Schuster, K. C., E. Urlaub, and J. R. Gapes. 2000. "Single-cell analysis of bacteria by raman microscopy: Spectral information on the chemical composition of cells and on the heterogeneity in a culture." *Journal of Microbiological Methods* 42 (1):29–38.

Sharma, G., and A. Prakash. 2014. "Combined use of fourier transform infrared and raman spectroscopy to study planktonic and biofilm cells of cronobacter sakazakii." *The Journal of Microbiology, Biotechnology and Food Sciences* 3 (4): 310–14.

Shi, Xianbo, Wei Yan, Dake Xu, Maocheng Yan, Chunguang Yang, Yiyin Shan, and Ke Yang. 2018. "Microbial corrosion resistance of a novel Cu-bearing pipeline steel." *Journal of Materials Science and Technology* 34 (12). Chinese Society of Metals: 2480–91.

Singh, J., and D. E. Wolfe. 2005. "Nano and macro-structured component fabrication by Electron Beam-Physical Vapor Deposition (EB-PVD)." *Journal of Materials Science.* Springer.

Suci, Peter A., Gill G. Geesey, and Bonnie J. Tyler. 2001. "Integration of Raman microscopy, differential interference contrast microscopy, and attenuated total reflection fourier transform infrared spectroscopy to investigate chlorhexidine spatial and temporal distribution in candida albicans biofilms." *Journal of Microbiological Methods* 46 (3). Elsevier: 193–208.

Tanaka, Keiichi, and Kazuo Maeda. 2014. "Scanning electron microscopy." *Journal of Health & Medical Informatics* 05 (04). OMICS International: 1000167.

Tiu, Brylee David B., and Rigoberto C. Advincula. 2015. "Polymeric corrosion inhibitors for the oil and gas industry: Design principles and mechanism." *Reactive and Functional Polymers.* Elsevier.

Usher, K. M., A. H. Kaksonen, I. Cole, and D. Marney. 2014. "Critical review: Microbially influenced corrosion of buried carbon steel pipes." *International Biodeterioration and Biodegradation.* Elsevier Ltd.

Vertes, Akos, Victoria Hitchins, and K. Scott Phillips. 2012. "Analytical challenges of microbial biofilms on medical devices." *Analytical Chemistry* 84 (9). American Chemical Society: 3858–66.

Wright, Chris J., Maia Kierann Shah, Lydia C. Powell, and Ian Armstrong. 2010. "Application of AFM from microbial cell to biofilm." *Scanning.*

Xu, Jin, Kaixiong Wang, Cheng Sun, Fuhui Wang, Ximing Li, Jiaxing Yang, and
    Changkun Yu. 2011. "The effects of sulfate reducing bacteria on corrosion of
    carbon steel Q235 under simulated disbonded coating by using electrochemical
    impedance spectroscopy." *Corrosion Science* 53 (4). Pergamon: 1554–62.
Zeiri, L., B. V. Bronk, Y. Shabtai, J. Czégé, and S. Efrima. 2002. "Silver metal induced
    surface enhanced Raman of bacteria." *Colloids and Surfaces A: Physicochemical
    and Engineering Aspects*, 208. Elsevier: 357–62.

6

Metallic Coatings

A. M. Kamalan Kirubaharan, K. Gobi Saravanan,
and Vinita Vishwakarma

CONTENTS

6.1 Introduction

Metallic coatings typically contain one or more metallic elements or alloys. Metallic coatings are mostly used for active and passive corrosion protection of metallic substrates. Metallic coatings are categorized into two main types depending on their electrochemical nature. One type provides corrosion protection (noble), and the other type not only provides a barrier to corrosive ions but also provides an active galvanic protection (sacrificial type). Zinc-based coatings are an example of the sacrificial type coating. Zinc deteriorates during service, defends the steel substrate by acting as a corrosion inhibitor, and provides electrons ($Zn \rightarrow Zn^{2+} + 2e^-$) to the steel substrate that arrests Fe^{2+} ions escaping from the steel. Another type is noble coatings, which act as a barrier between the metal substrate and the surroundings. Copper, lead, chromium, nickel, and silver-based coatings fall under this category of noble metal coatings. This type of metallic coatings is effective only when the coating is free from pores or cracks. The formation of cracks and pores in the noble metal coating causes galvanic corrosion.

An important function of metallic coatings is that the coating can significantly influence the corrosion behavior of the coated metal components. Generally, the application of metallic coatings can be done using a sprayer, chemically, mechanically, or electrochemically. Such metallic coatings are generally applied on metallic substrates to provide glossy or shiny appearances and protect from sunlight, oxidation, and corrosion. The presence of defects in a metallic coating causes damage to the underlying substrate through the formation of local cells.

Metallic pipelines are widely used in transporting gas, oil, water, and steam. Corrosion in pipelines occurs due to contact with the external environment causing initial rusting followed by corrosion. Corrosion in pipelines shortens their reliability and their functionality. Therefore, the service life of these pipelines is highly threatened by factors like corrosion or chemical deterioration. Such factors can end up affecting the activities of industries wherein pipelines have a major usage. To avoid such occurrences, coatings for pipelines for their protection from corrosion are extremely essential.

Metallic coatings offer an economical way to control or modify the properties of a material. Another significant use of metallic coatings is to offer wear and tear resistance. The main criterion for the use of metallic coating is to provide a strong adhesion with the base material. Most of the coatings are applied through the hot dipping process, such as aluminizing or galvanizing. The formation of a thin/thick coating of a more reactive metal upon a less reactive material (e.g., Fe-based alloy) can act as a sacrificial coating. Electro and electroless plating are the most widely used methods of applying metallic coatings. Other plating processes such as hot dipping, chemical vapor disposition (CVD), air plasma spraying (APS), RF/DC magnetron sputtering, and cementation have also been used. Each technique has its own advantages and the right technique is selected depending on the operational conditions and the nature of application.

Metallic coatings generally consist of one or more pure metals such as nickel (Ni), cadmium (Cd), zinc (Zn), tungsten (W), cobalt (Co), phosphorus (P), copper (Cu), and iron (Fe) (Abdeen et al. 2019). A metallic coating can be of a metal or an alloy for the enhancement of corrosion resistance. The corrosion performance of pure zinc is inadequate under severe oxidizing conditions; hence, it is alloyed to provide a higher corrosion resistance than a metallic coating. Alloyed metallic coatings can provide superior barrier properties to that of a pure metal. For example, Zn–Ni alloy plating has shown better hardness than a single layer of zinc and cadmium coatings (Youssef, Koch, and Fedkiw 2004).

In addition to that, active corrosion inhibitors can be incorporated along with the addition of some of the alloying elements. Among the possible replacements for chromium, rare-earth elements and transition metal (TM) ions (Ce^{3+}, Co^{2+}, and MoO_4^{2-}) are possible selections to prevent localized corrosion of alloys and steels. For example, transition and rare-earth metals (such as Fe, Ni, Ce, Co, Gd, and Y) are added to alloy or metallic coating up to 15 at.% to improve the amorphous nature of the coating (Moreno et al. 2008). Metals such as Fe, Co, and Ni are added to increase the localized corrosion resistance, whereas Y^{3+}, Ce^{3+}, and Gd^{3+} inhibit corrosion in aqueous solutions (Kabacoff et al. 1991).

6.2 Techniques of Metallic Coatings and Significance

6.2.1 Hot Dip Galvanizing

Hot-dip galvanizing was well known to the world from the beginning of the 17th century. It is vital for steel and iron for protecting their life and utility from environmental corrosion. The most widely used single layer of metallic coating in view of volume of metal used is through the hot dip galvanizing technique. In this method, zinc is plated to steel and iron parts by immersing the components into a bath containing molten zinc. In the bath, the molten zinc metallurgically reacts with iron in the steel to provide a strongly bonded uniform alloy coating, which provides excellent corrosion protection to the steel substrate. Aluminum and cadmium are the only types of unique combinations. Lead (Pb) is often used in the molten zinc bath in order to improve the viscosity of the bath. A hot-dip galvanized coating is relatively cheaper and easier than an organic-based paint coating that provides equivalent corrosion protection. The durability of a hot-dip galvanized coating solely depends on the corrosion rate with which it is being placed.

6.2.2 Thermal Spray Coating

Thermal spraying or plasma spraying of metal and oxide ceramic coatings has numerous properties that are suitable for applications including corrosion and wear resistance. In this process, the molten materials are plasma sprayed over the surface to be coated. The types of thermal spray processes include plasma arc spray, laser spray, electric arc spray, and flame spray depending on the power source used. Thermal spraying can be used to develop thick coatings (20 μm to a few mm) depending on the feedstock. This process has a higher deposition rate than other coating processes such as electro/electroless plating, chemical vapor deposition, and physical vapor deposition.

The potential benefits of the plasma spraying process are summarized below:

1. It can be applied both manually and automatically.
2. By selecting the proper choice of material, the durability of the coated materials can be extended.

3. Can spray materials having a higher melting point (including ceramic materials).

4. Thick/thin coatings can be deposited at higher rates.

5. A wide range of materials can be coated (e.g., metals, alloys, carbides, and ceramics).

The type of bonding between plasma-sprayed coating and substrate is through mechanical and not metallurgical bonding. The adhesion to the substrate depends on the substrate surface, which needs to be cleaned and abraded by grit blasting prior to thermal spraying. This type of thermal spray process is used to develop chromium carbide (Cr_3C_2) coatings or erosion-resistant nickel-chromium (Ni-Cr), copper-nickel-indium (Cu-Ni-In), and wear and abrasion-resistant tungsten carbide-cobalt coatings.

6.2.3 Thermal Evaporation

Thermal evaporation is one of the common types of physical vapor deposition (PVD) technique. This is one of the simple techniques in PVD coatings and uses a resistive type heat source to evaporate a metal (melting point of the material should be less than ~1,450°C) under a vacuum environment to develop a coating. The material is heated in a vacuum chamber till the vapor pressure is produced and condensed onto the substrates. It is one of the most widely employed processes for the deposition of alloys, metals, and also many compounds.

The source material from thermal evaporation condenses at the substrate with gas molecules between the source and substrate holder. Thermal evaporation is a 'line-of-sight' process. The vacuum condition additionally helps to reduce the contamination in the deposition chamber to a low level. Normally, thermal evaporation carried out at a gas pressure of ~10^{-6} Torr relies upon the level of vaporous contamination that can be endured in the deposition system. In most cases, the substrates are mounted in the substrate holder at a distance away from the source to reduce radiation heating of the substrate by the vaporization source. Thermal evaporation is commonly used to develop metallic coatings on a metal substrate. Few advantages include low cost, high deposition rates, and relative simplicity, of the equipment.

6.2.4 Sputtering

Sputtering is another type of PVD method used to deposit thin films/coatings of a metallic material onto a substrate's surface, by generating gaseous plasma and ions (Figure 6.1).

Sputtering can be used for both conductive and non-conductive materials. This technique gives excellent adhesion to the substrate and uniform coatings than other vacuum deposition techniques. This technique can be

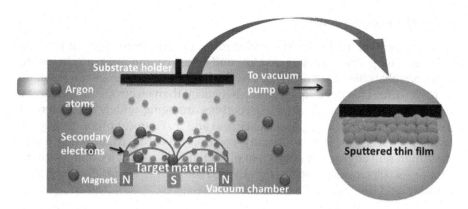

FIGURE 6.1
Schematic diagram of the sputter deposition technique.

utilized in a range of materials, including metal alloys and oxides. The equipment, operating, and maintenance costs are higher in this type of deposition, but the deposition rate is lower. The advantage of sputtering is that extremely high melting point materials can be easily coated. The composition of sputtered metal/alloy coatings is close to that of the source material.

6.2.5 Electroplating

The electroplating technique has expanded rapidly. Electroplating is a process of metallic coating over another metal with the help of an electric current passed through a chemical solution. This electroplating process produces a thin/thick metallic coating onto the surface of the metal substrate. The negatively charged object to be coated is submerged into a solution of positively charged metal ions. When the positively charged ions meet the negatively charged object, they become neutral and hence plating occurs. Since the ions are no longer charged, they settle down as a very thin coat on the object to be plated.

The most commonly plated metal is zinc, followed by chromium, nickel, and copper. Several metals and metalloids can be electrodeposited (such as Mn, Fe, Co, Ga, Ge, As, Se, Ru, Cd, In, Sn, Pb, Bi, Hg, Rh, Pd, Ag, Sb, Au, Ir, and Pt). Chromium electroplating is biologically harmless due to its inertness, but hexavalent chromium used in electroplating baths is toxic as well as a suspected carcinogen. The removal of cyanides from the electroplating baths has become a goal since the 1980s.

The purpose of electroplating is to improve the appearance and protection of the metal against corrosion. Electroplating is most widely used in industries to improve the surface qualities of metallic components such as abrasion resistance, corrosion, reflectivity, lubricity, electrical conductivity, and appearance. In recent years, the advancements in the electroplating technology result in improved coating functionality, economy, and need for environmental and judicial compliance.

6.2.6 Electroless Plating

Brenner and Riddell invented the electroless coating technology in the year 1946. In this process, metals, oxides, and other compounds are deposited onto the substrates as a thin layer. The thickness of the coatings can reach up to 10–200 μm. This technique is more beneficial than electroplating (Sharma, Cheon, and Jung 2016). Electroless coatings are being widely used increasingly due to their unique physicochemical and mechanical properties. Some of the properties include uniformity, corrosion/wear resistance, high hardness, high reflectivity, low coefficient of friction and resistivity, and magnetic properties. Electroless coatings are divided into three categories: (i) metallic, (ii) alloy, and (iii) composite coatings.

6.2.6.1 Electroless Nickel Plating

Electroless nickel plating is a process of plating onto a metallic surface through an autocatalytic chemical reaction. Unlike the electricity sources used from outside (electroplating), electroless nickel plating uses a chemical bath to produce a nickel layer onto metallic substrates. Electroless nickel can be coated over non-conductive surfaces also, which allows this technique to be used for a wider variety of base materials. Electroless nickel plating shows better corrosion resistance and provides a uniform coating with high precision. This can be coated to both ferrous and non-ferrous surfaces of any complex shape dimensions. The unique advantages in this process include corrosion/wear and friction resistance.

The presence of pores in the electroless nickel plating is very less and provides good corrosion resistance to steel substrates. This plating can be developed with zero or less compressive stress, making it good for end-use. The main advantage of this technique is that no electricity is used externally to start this process, which makes it more efficient and precise in a cost-effective way. This process can be done with less manpower, equipment, and lesser coats required than electroplating to make it a more durable and high-quality finish. This technique also provides large flexibility in thickness to volume of the plating and fills the pits easily on the substrates of the material to be coated. This makes it suitable for a wide variety of industrial components that can be coated with a uniform surface finish including valve pumps, pipelines, electrical, mechanical tools, and engineering components.

Different types of nickel can be used for plating, which contains phosphorus in different concentrations (low, medium, and high) depending on the requirement. Most of this coating process uses a medium level of phosphorus content. Such coatings contain phosphorus content that falls between 5% and 9%. Such type of electroless nickel plating has higher deposition rates as well as controls the brightness of plating from full bright to semi options. Phosphorus content less than 5% provides uniform thickness and eliminates the grinding process after plating and also improves the alkaline resistance to corrosion. Higher concentrations of phosphorus content between 10% and 13% in nickel electroplating are used in coal mining and oil drilling industries where a higher amount of corrosion resistance is required from highly corrosive acids.

6.2.6.2 Electroless Ni-P Coatings

Electroless Ni-P coatings are commonly employed to enhance the surface properties of metallic structures in several engineering applications mainly for aerospace and chemical industries. Tailoring of electroless nickel-based coatings' surface property is predominantly useful for applications in the gas and oil industries. Steel pipelines suffer from erosion-enhanced corrosion issues that reduce their durability (Wang et al. 2017).

In recent years, researchers have paid huge attention to electroless plating of nickel-based ternary alloys because of their superior wear, corrosion, thermal stability, and electrical properties (Georgieva and Armyanov 2007). For example, the addition of Cu in the Ni-P coating offers a smooth surface, brightness, and corrosion resistance. The best corrosion resistance behavior is observed in a Ni–Sn–P-based coating and shows improved solderability than a Ni-P coating.

6.2.6.3 Electroless Ternary Composite Coating

In ternary electroless plating, the desired metal compounds are added into an electrochemical bath to provide the metal ion deoxidizer and to coat onto the substrate, which elevates to more significant accomplishments in electroless plating methods. Aal et al. developed a Ni–Cu–P coating from a Ni–P bath containing $CuSO_4.5H_2O$, which is coated over stainless-steel foams for heat exchangers and filtration applications. From the study, it is illustrated that a low rate of deposition, finer grain sizes, and enhanced corrosion resistance were observed by increasing the Cu content into the Ni–P matrix (Abdel Aal and Shehata Aly 2009). Electroless Ni–Mo–P plating on steel in 0.5 mol/L H_2SO_4 exhibits superior corrosion resistance than Ni-P coating (Song et al. 2017). Since tungsten metal (W) shows versatile properties such as a higher melting stage, hardness and higher stretching intensity, Ni–W–P electroless coatings have been performed on stainless steel (Wu et al. 2004), mild steel (Tien, Duh, and Chen 2004), and tool steel surfaces. The abrasion resistance and hardness were improved significantly in electroless plated samples at elevated post heat treatment. Wang et al. observed that a double-layer Ni–Cr–P composite coating exhibited higher corrosion resistance than Ni–P alloy electroless coating (Wang et al. 2017). Wang, Ji, and Lee (2013) developed Ni–Fe–P–B and Ni–Fe–P alloy deposits in the presence of an alkaline solution containing KBH_4 or hypophosphite as a reducing agent with controlled concentrations of KBH_4 and NaH_2PO_2. From the investigation, it is illustrated that KBH_4 causes less corrosion resistance in an alkaline environment, but the corrosion resistance nature was slightly improved on Ni–Fe–P deposits.

Electroless plating is a kind of surface modification technique, in which chemical reduction of a metallic ion will take place in an aqueous solution comprising a stabilizing agent and a reducing agent. In this technique, subsequent deposition of the metal is performed without electrical energy, especially differing from the electroplating technology. The formation of Ni-P coatings by the electroless deposition technique has been a recognized commercial process that exhibits numerous applications in various fields because of outstanding properties of coatings, such as high wear and corrosion resistance, high hardness, acceptable lubricity, good lubricity, and superior compatibility on different suitable substrates such as aluminum alloy, steel, copper, ceramics, magnesium alloy, and polymers with different needs of applications.

6.3 Corrosion and Leaching Problems

Metallic corrosion is the deterioration of metals by a redox process in which metals are oxidized in the presence of moisture. Examples include iron-rusting, oxidation of silver, and the formation of a green film on copper and brass. Corrosion in metallic coatings is an electrochemical process. In an aqueous solution, current flows in the metal from reduction sites to oxidation sites. An ionic current flow in the solution has completed the circuit (electrolyte). In atmospheric corrosion, the electrolyte is thinner or discontinuous, or both. The formation of a nanometer thick passivation layer from metals often intervenes the corrosion rate.

By performing the surface modification of a material through metallic coating/plating, the material can withstand harsh environmental conditions. Corrosion in metal can occur by several different mechanisms: (i) pitting, (ii) galvanic corrosion, (iii) inter-granular attack, (iv) leaching, (v) corrosion fatigue, (vi) stress corrosion cracking (SCC), and (vii) erosion. Galvanic corrosion is a very common type of corrosion and occurs when two dissimilar metals get in contact or are welded to provide a conductive path. When a coated material is subjected to extreme thermal stress, a material may experience corrosion (SCC) along the grains, which leads to corrosion in the localized region. Pitting corrosion arises when a localized part of a material experiences corrosion at a much faster rate than other parts of the component.

To protect galvanic corrosion in iron and steel, zinc and aluminum coatings are used. Larger components, such as pipelines and windmills, are often coated with aluminum and zinc-based coatings because they provide stable and durable corrosion resistance. Zinc-coated steel is a typical example of sacrificial anode type. The zinc corrodes and protects the underlying steel substrates over which it is coated. The lifetime of the coating has been determined from the thickness of the coating to be deposited. Iron and steel fasteners are repeatedly coated with a layer of cadmium (Cd), which blocks the hydrogen absorption process and thus leads to stress corrosion cracking (Barrera et al. 2018). Cobalt-chromium (Co-Cr) (Chisholm 1975) and nickel-chromium (Ni-Cr) (Kunyarong and Fakpan 2018) coatings are used as corrosion-resistant coatings due to the presence of less level of porosity. These coatings show exceptional moisture resistance and help to inhibit the formation of rust and further deterioration of the metal. Coatings containing chromium and phosphorus can act as effective corrosion inhibitors.

6.4 Industrial Applications

Nickel and other metals can act as a sacrificial coating that resists rust, which is an important requirement for offshore oil and gas operations. Zinc coating is preferred for use in nuts, bolts, and brackets as it produces clear surface finishing and slows down the rusting and other corrosion processes. In addition, zinc is also frequently used as a base coating (primer) on surfaces prior to painting. Coatings made up of zinc alloys are widely used in the automobile industry for several years. Other uses include chemical carrying pipelines, heavy electrical components, and manufacturing of tanks and military personnel carriers. Some types of metal coatings are designed to protect metals from dirt, rust, debris, and corrosion. Such types of coatings are substantial for applications in outdoor equipment, like boats, automobiles, trains, heavy equipment, and airplanes.

References

Abdeen, Dana H., Mohamad El Hachach, Muammer Koc, and Muataz A. Atieh. 2019. "A review on the corrosion behaviour of nanocoatings on metallic substrates." *Materials* 12 (2): 1–42.

Abdel Aal, A., and M. Shehata Aly. 2009. "Electroless Ni-Cu-P plating onto open cell stainless steel foam." *Applied Surface Science* 255 (13–14). Elsevier: 6652–55.

Barrera, O., D. Bombac, Y. Chen, T. D. Daff, E. Galindo-Nava, P. Gong, D. Haley, et al. 2018. "Understanding and mitigating hydrogen embrittlement of steels: A review of experimental, modelling and design progress from atomistic to continuum." *Journal of Materials Science* 53 (9): 6251–90.

Chisholm, C. U. 1975. "Cobalt-chromium coatings by electrodeposition: Review and initial experimental studies." *Electrodeposition and Surface Treatments* 3 (5–6). Elsevier: 321–33.

Georgieva, J., and S. Armyanov. 2007. "Electroless deposition and some properties of Ni-Cu-P and Ni-Sn-P coatings." *Journal of Solid State Electrochemistry* 11 (7): 869–76.

Kabacoff, L. T., C. P. Wong, N. L. Guthrie, and S. Dallek. 1991. "Formation and stability of magnetron sputtered Al-TM-RE metallic glasses." *Materials Science and Engineering A* 134 (C). Elsevier: 1288–91.

Kunyarong, Aumpava, and Kittichai Fakpan. 2018. "Cr-Ni alloy coating electrodeposited on T22 steel." *Materials Today: Proceedings*, 5. Elsevier Ltd.: 9244–49.

Moreno, Presuel F., M. A. Jakab, N. Tailleart, M. Goldman, and J. R. Scully. 2008. "Corrosion-resistant metallic coatings." *Materials Today* 11 (10): 14–23.

Sharma, Ashutosh, Chu-Seon Cheon, and Jae Pil Jung. 2016. "Recent progress in electroless plating of copper." *Journal of the Microelectronics and Packaging Society* 23 (4): 1–6.

Song, Gong Sheng, Shuo Sun, Zhong Chi Wang, Cheng Zhi Luo, and Chun Xu Pan. 2017. "Synthesis and characterization of electroless Ni-P/Ni-Mo-P duplex coating with different thickness combinations." *Acta Metallurgica Sinica (English Letters)* 30 (10). The Chinese Society for Metals: 1008–16.

Tien, Shih Kang, Jenq Gong Duh, and Yung I. Chen. 2004. "Structure, thermal stability and mechanical properties of electroless Ni-P-W alloy coatings during cycle test." *Surface and Coatings Technology* 177–178 (January). Elsevier: 532–36.

Wang, Chuhong, Zoheir Farhat, George Jarjoura, Mohamed K. Hassan, and Aboubakr M. Abdullah. 2017. "Indentation and erosion behavior of electroless Ni-P coating on pipeline steel." *Wear* 376–377 (April). Elsevier Ltd: 1630–39.

Wang, Qin Ying, Yu Chen Xi, Jiao Xu, Shuang Liu, Yuan Hua Lin, Yun Hong Zhao, and Shu Lin Bai. 2017. "Study on properties of double-layered Ni–P–Cr composite coating prepared by the combination of electroless plating and pack cementation." *Journal of Alloys and Compounds* 729 (December). Elsevier Ltd: 787–95.

Wang, Wei, Shaowen Ji, and Ilsoon Lee. 2013. "A facile method of nickel electroless deposition on various neutral hydrophobic polymer surfaces." *Applied Surface Science* 283 (October). Elsevier B.V.: 309–20.

Wu, F. B., S. K. Tien, W. Y. Chen, and J. G. Duh. 2004. "Microstructure evaluation and strengthening mechanism of Ni-P-W alloy coatings." *Surface and Coatings Technology* 177–178 (January). Elsevier: 312–16.

Youssef, Kh.M.S, C C Koch, and P S Fedkiw. 2004. "Improved corrosion behavior of nanocrystalline zinc produced by pulse-current electrodeposition." *Corrosion Science* 46 (1): 51–64.

7

Ethical Issues and Environmental Safety

Vinita Vishwakarma and A. M. Kamalan Kirubaharan

CONTENTS

7.1 Introduction

In recent years, the awareness of ethical issues and environmental safety toward the coating technology for the industrial application is considered. Deterioration of metals and materials without suitable protective layers is very common in harsh environments. Many hazardous materials are used during the coating process and several ethical issues and environmental safety are involved. The usages of toxic chemicals have reverse effects on the workers of the factory. Before using the chemicals for coatings, the detailed chemical instructions should be learned by the employee, who is going to perform the coating task. More number of chemicals such as solvents, additives, and pigments are used by the coating industries on a daily basis and workers are exposed to those chemicals. These chemicals develop health risks and genetic disorders (chromosomal aberration) as well as environmental threats. While performing the coating process, the hazardous chemicals enter into the human body via different routes and affect the respiratory system, liver, kidney, reproductive system, brain, etc. As required, the employee should be alert for the safe handling of the coating process and emergency control measures. Pipeline coating is significant in preventing the pipes from biofouling, biocorrosion, and deterioration. Apart from organic and inorganic materials, nanomaterials have a new perception in coating application. The safety issues of these nanomaterials need to be concentrated from the safety point of view to protect the workers during the coating process. It is also important to control the release of coatings to protect the environment.

7.2 Safety and Ethics of Metallic Coatings

The metallic coatings are essential for the industrial components and to protect the materials from the threat of biofouling and corrosion. Coating of pipeline materials is important to guard them against the different environmental conditions and to enhance their efficiency and workability. Most of the time, these coatings are a risk from a health and safety point of view for the industrial workers who are exposed to these hazardous coating chemicals and have chances of health problems like muscle fragility, memory loss, headache, lack of sleep, and so on. The coating chemicals used in conventional organic coating processes such as dip coating, brushing, rolling, spray, spin coating, and so on are dangerous and pose serious health risks for the workers. In the coating process, mostly heavy metals are used such as zinc, cobalt, nickel, chromium, and lead. Some organic solvents such as alcohols, hydrocarbons, glycols, ketones, and esters are used, which are carcinogenic, flammable, and explosive. The risk of injury from the surrounding areas is also a threat for the workers. The workers must be aware of the safety instructions of the safe handling of the coatings and emergency control measures. The deterioration inside the pipeline is because of corrosive components and the toxic chemical coating inside the pipeline is a health risk for the workers.

The inorganic chemical coatings are more ideal than organic coatings. The powder metallic coating is used by many industries from the safety point of view considering that there will be no harmful effect and it is environmentally friendly. The significance of pipeline coating is to protect pipelines from cracks, corrosion, and biofouling and to enhance their life. Nanomaterials have a new perspective in the coating industries. Workers when working with nanomaterials for metallic coatings must take care of the extra safety ethics for their health, because these materials with their high surface areas are easily absorbed by the body through inhalation, dermal exposure, and ingestion compared to materials with lower surface areas (Gajewicz et al. 2012). It has also been reported that these nanoparticles penetrate the cell membrane and are a reason for cell abnormalities and death (Colvin 2003; Osman 2019). The safety issues of nanomaterials need to be addressed by the Environmental Health and Safety (EHS) committee team of the industries' personnel and guide the workers. Before handling any chemicals, the workers should follow the instructions provided in the Materials Safety Data Sheet (MSDS). Metal oxide nanoparticles such as SiO_2, TiO_2, ZnO_2, Fe_2O_3, and so on are used as coating materials by the industries, which exhibit distinct toxicities under various environments. From the safety point of view, green approaches to prepare the metal nanoparticles for coating application will be effective. So, to protect the workers, coating industries must implement novel coating techniques to reduce the hazardous coating materials for the safety of the workers.

7.3 Controlled Release of Antifouling Coatings

Biofouling is a universal problem associated with the attachment on any substrate. To prevent biological adherence, the antifouling coatings are an effective and easy method on the substrate. Unfortunately, the materials used for antifouling have toxic molecules because of their toxic nature such as acrylic resin tributyltin (TBT), copper, and so on (Lagerström et al. 2017). Therefore, it is essential to analyze the release rate of fouling coating in a controlled manner based on the time duration and develop an environmentally friendly and non-toxic coating on any substrate (Hu et al. 2020; Barletta et al. 2018; Turner 2010). The release rate of coated materials depends on the method of synthesis and selection of the materials and optimized parameters such as pH, temperature, coating time, water solubility, thickness of the film, adhesion, and exposed condition of the samples. The coating chemicals such as hydrogen peroxide are environmentally friendly as they decompose quickly without any effect on the environment and animals (Olsen et al. 2010).

The antifouling coating by epoxy polymer has shown an excellent controlled release rate with the eco-friendly biocide based on the hydrophilic, hydrophobic, and biodegradable approaches (Ali et al. 2020). The hydrophilic and hydrophobic coating avoids the attachment of marine organisms to the ship hull (Wan et al. 2012). Bilayer and multilayer metallic coatings by copper (Cu), silver (Ag), zinc (Zn), nickel (Ni), chromium (Cr), etc. control the release of metallic ions into the environment (Vishwakarma et al. 2009). Metal oxide nanoparticles have more durability against the microbial attack which prevents the substrate from the fouling attack (Zhang et al. 2016; Nurioglu, Esteves, and De With 2015; Cao et al. 2011). The biocides such as sodium hypochlorite (NaOCl) coating initially inhibit the slime formation of the biofouling but to prevent the microbial-induced corrosion (MIC) in the pipeline, shock dosing of two types of biocides is essential to remove it. Biocides are toxic so their use should be minimized for the controlled release into the environment (Edge et al. 2001). It has been found that the selection of antifouling coating materials is controlled by the compatibility of the materials along with the desired properties and environmental conditions (Barroso et al. 2019). The synthesized materials for coating should be environmentally friendly with a controlled release of leaching of the materials.

References

Ali, Abid, Muhammad Imran Jamil, Jingxian Jiang, Muhammad Shoaib, Bilal Ul Amin, Shengzhe Luo, Xiaoli Zhan, Fengqiu Chen, and Qinghua Zhang. 2020. "An overview of controlled-biocide-release coating based on polymer resin for marine antifouling applications." *Journal of Polymer Research*, 27(4): 1–17. Springer.

Barletta, M., C. Aversa, E. Pizzi, M. Puopolo, and S. Vesco. 2018. "Design, development and first validation of 'Biocide-Free' anti-fouling coatings." *Progress in Organic Coatings* 123 (October). Elsevier B.V.: 35–46.

Barroso, Gilvan, Quan Li, Rajendra K. Bordia, and Günter Motz. 2019. "Polymeric and ceramic silicon-based coatings-a review." *Journal of Materials Chemistry A*, 7(5): 1936–1963. Royal Society of Chemistry.

Cao, Shan, Jia Dao Wang, Hao Sheng Chen, and Da Rong Chen. 2011. "Progress of marine biofouling and antifouling technologies." *Chinese Science Bulletin*, 56(7): 598–612. Springer.

Colvin, Vicki L. 2003. "The potential environmental impact of engineered nanomaterials." *Nature Biotechnology*, 21(10): 1166–70. Nature Publishing Group.

Edge, M., N. S. Allen, D. Turner, J. Robinson, and Ken Seal. 2001. "The enhanced performance of biocidal additives in paints and coatings." *Progress in Organic Coatings* 43 (1–3). Elsevier: 10–17.

Gajewicz, Agnieszka, Bakhtiyor Rasulev, Tandabany C. Dinadayalane, Piotr Urbaszek, Tomasz Puzyn, Danuta Leszczynska, and Jerzy Leszczynski. 2012. "Advancing risk assessment of engineered nanomaterials: Application of computational approaches." *Advanced Drug Delivery Reviews*, 64(15): 1663–1693. Elsevier.

Hu, Jiankun, Baoku Sun, Haichun Zhang, Ading Lu, Huiqiu Zhang, and Hailong Zhang. 2020. "Terpolymer resin containing bioinspired borneol and controlled release of camphor: Synthesis and antifouling coating application." *Scientific Reports* 10 (1). Nature Research: 1–10.

Lagerström, Maria, Jakob Strand, Britta Eklund, and Erik Ytreberg. 2017. "Total tin and organotin speciation in historic layers of antifouling paint on leisure boat hulls." *Environmental Pollution* 220 (January). Elsevier Ltd: 1333–41.

Nurioglu, Ayda G., A. Catarina C. Esteves, and Gijsbertus De With. 2015. "Non-toxic, non-biocide-release antifouling coatings based on molecular structure design for marine applications." *Journal of Materials Chemistry B*, 3 (32), 6547–6570. Royal Society of Chemistry.

Olsen, S. M., J. B. Kristensen, B. S. Laursen, L. T. Pedersen, K. Dam-Johansen, and S. Kiil. 2010. "Antifouling effect of hydrogen peroxide release from enzymatic marine coatings: Exposure testing under equatorial and mediterranean conditions." *Progress in Organic Coatings* 68 (3). Elsevier: 248–57.

Osman, Eman M. 2019. "Environmental and health safety considerations of nanotechnology: Nano safety." *Biomedical Journal of Scientific & Technical Research* 19 (4): 14501–14515. Biomedical Research Network, LLC.

Turner, Andrew. 2010. "Marine pollution from antifouling paint particles." *Marine Pollution Bulletin* 60 (2). Pergamon: 159–71.

Vishwakarma, Vinita, J. Josephine, R. P. George, R. Krishnan, S. Dash, M. Kamruddin, S. Kalavathi, N. Manoharan, A. K. Tyagi, and R. K. Dayal. 2009. "Antibacterial copper-nickel bilayers and multilayer coatings by pulsed laser deposition on titanium." *Biofouling* 25 (8). Taylor & Francis: 705–10.

Wan, Fei, Xiaowei Pei, Bo Yu, Qian Ye, Feng Zhou, and Qunji Xue. 2012. "Grafting polymer brushes on biomimetic structural surfaces for anti-algae fouling and foul release." *ACS Applied Materials and Interfaces* 4 (9). American Chemical Society: 4557–65.

Zhang, Yang, Ying Wan, Yuanteng Shi, Guoyuan Pan, Hao Yan, Jian Xu, Min Guo, Longxin Qin, and Yiqun Liu. 2016. "Facile modification of thin-film composite nanofiltration membrane with silver nanoparticles for anti-biofouling." *Journal of Polymer Research* 23 (5). Springer Netherlands: 1–9.

8

Future Prospects of the Coating Technology

Vinita Vishwakarma and Dawn S S

CONTENTS

8.1 Introduction

Coating technology is an essential part of the industries globally due to its ability to protect materials and its cost effectiveness. Biofouling, biocorrosion, and biodeterioration are the major challenges of the materials especially in the pipelines, where surface modification of the materials plays a significant role. The protective coatings of the metals and materials are of great interest for the industrial application to enhance their structural, functional, and esthetic properties. The global market for the industrial coating will grow from $158.6 to 203.8 billion commencing from 2021 to 2026. The upcoming smart coatings to protect the materials from biofouling, biocorrosion, and biodeterioration are still in their infancy. Materials such as stainless steel (SS), titanium (Ti), aluminum (Al), carbon steel (CS), and so on are very much susceptible to corrosion, pitting, microbial attack, etc. There are various conventional methods to perform the coating on the substrate materials. There was a traditional solvent-borne coating which has transformed into polymer coating, solid, and powder coatings. Solvent-borne epoxy, acrylic, polyurethane (PU), vinyl, and polyester coatings are conventionally used for protecting the materials from chemical and microbial corrosion. PUs are an important class of polymers that have found many applications starting from coating, paint, foam, thermosetting, and thermoplastic elastomer to fiber (Thakur and Karak 2013).

Commercially, petrochemical-based polyols are used in PUs, but with the escalating price of crude petroleum and environmental concern, the demand for vegetable oil-based polyols is growing. The development of eco-friendly PU coatings based on neem oil was evaluated for gloss, scratch hardness, adhesion, flexibility, thermal stability, impact, and chemical composition (Chaudhari et al. 2013). Vegetable oils such as castor oil (*Ricinus communis*), karanja (*Pongamia glabra*), and sugar-apples (*Annona squamosa*) have been reported for corrosion inhibition in steel in acid media (Srivastava and Srivastava 1981; Sathiyanathan et al. 2005; Raja and Sethuraman 2008). Vegetable oil from seeds of several edible and medicinal plants has also been employed for anticorrosion behavior against alkaline and acid media as well as chloride ions (Abdullah Dar 2011; Lahhit et al. 2011). Recent environmental awareness and depletion of world fossil fuel reserves have motivated us to look for a substitute for petroleum-based PU coating. At present, vegetable oils and their derivatives including fatty acids, fatty acid esters, and crude glycerol have shown good potential in producing bio-based polyols and PUs. Despite their promise, virgin vegetable oil is expensive to synthesize PUs (Enderus and Tahir 2017).

Vegetable oils and their derivatives such as fatty acids, fatty acid esters, and crude glycerol have good potential as renewable and sustainable feedstock in producing bio-based polyols and PUs (Li, Luo, and Hu 2015). Vegetable oils are used in various industrial applications such as emulsifiers, lubricants, plasticizers, surfactants, plastics, solvents, and resins (Erhan 2005). Because of their high versatility, polyols from vegetable oils and their derivatives have been used to produce various PU materials such as foams, elastomers, rigid plastics, and coatings, which have shown properties mostly comparable to those of their petroleum-based analogs (Li, Luo, and Hu 2015).

Polyols are produced from epoxidized vegetable oil by oxirane ring-opening reactions using a broad range of active hydrogen-containing compounds such as alcohols, inorganic and organic acids, amines, water, and hydrogen (Pan and Webster 2012; Miao et al. 2010; Dai et al. 2009; Guo, Cho, and Petrović 2000; Kiatsimkul et al. 2008; Wang et al. 2009; Caillol et al. 2012). Vegetable oil-based polymers are used for the development of paints and coatings as they are non-toxic, non-depletable, domestically abundant, non-volatile, and biodegradable (Xia and Larock 2010; Salimon, Salih, and Yousif 2012). A series of PUs from polyols derived from soybean, corn, safflower, sunflower, peanut, olive, canola, and castor oil were prepared, and their thermal stability in air and nitrogen was assessed by thermogravimetric analysis, FTIR, and GC/MS (Javni et al. 2000). A comparative study was carried out with standard petro-based polyols along with *Gossypium arboreum* plant oil and dicarboxylic acids as newer eco-friendly raw materials for PU coatings (Patil et al. 2017). The role of castor oil was found in epoxy and polyamide systems for coating and adhesive application (Shukla et al. 2005). PU coatings based on 100% glycerolysis-based polyols were prepared from linseed, soybean, and sesame oils. The produced coatings generally showed satisfactory flexibility and adhesion properties as well as good chemical resistance (Gite et al. 2006).

8.2 Prospects of the Coating Technology

Vegetable oils and their derivatives have been widely used to produce polyols and PUs owing to their low toxicity, unique triglyceride structure, thermal stability, low cost, and biodegradability. Vegetable oils such as castor, tung {wood}, corn, olive, canola, soybean, rapeseed, sunflower, palm, linseed, and passion fruit oil have been used for PU production. Earlier reports show castor oil with a higher content of ricinoleic acid (87%–90%) with a secondary hydroxyl (OH) functional group and less edibility to be a promising source for bio-based PU synthesis. PU exhibits a higher antifouling performance even in a static marine environment (Xu et al. 2014). The development of corrosion-resistant epoxy coating for pipelines improves the life and efficiency of the system where the multilayer coatings are more effective for anticorrosion compared to a single layer with the addition of nanoparticles (Buhri, Kaithari, and Rasu 2016). PU is also used and exhibits a higher antifouling performance even in a static marine environment (Xu et al. 2014). However, waste cooking oils (WCO)-based polyols for preparing PU coatings have not been explored intensely.

Owing to the edible aspects of other vegetable oils and to avoid the competition of resources, WCO will be explored as raw materials to synthesize PUs. The PUs synthesized from WCO, which is a vegetable oil instead of virgin vegetable oil, have similarities in their organic structures that make WCO suitable as the source to synthesize polyols using the same reaction as used to produce virgin vegetable oil-based polyols (Ullah, Bustam, and Man 2014; Mohd Tahir et al. 2016; Enderus and Tahir 2017). The pretreated WCO by using coconut husk-activated carbon followed by the transesterification reaction to synthesize polyol has been prepared (Enderus and Tahir 2017). Comparing the pathways for synthesizing vegetable oil-based polyols that take place at the double bond moieties of vegetable oils, the conversion of vegetable oils to produce polyols via the transesterification pathway takes place at the ester moieties in the vegetable oil structures. Pentaerythritol (Chuayjuljit, Maungchareon, and Saravari 2010), triethanolamine (Stirna, Cabulis, and Beverte 2008), and glycerol (Mohd Tahir et al. 2016) are popularly used alcohols in polyol synthesis by transesterification. It was explained that PU products for coatings and adhesives were also developed from WCO (Panadare and Rathod 2015). In India, not much work has been done on the WCO-based PU coatings and their application in pipelines.

To enhance the durability of the materials, prevent corrosion, and enable self-cleaning and antibacterial properties of the waste cooking oil-based PUs, nanoparticles that are suitable for enhancing the coating properties must be incorporated. Polymers containing TiO_2 nanoparticles have attracted significant attention due to their corrosion resistance and biocidal properties (Toker, Kayaman-Apohan, and Kahraman 2013; Guo et al. 2013), and self-cleaning properties due to the photocatalytic activity of TiO_2 (Cassar 2005). In the coating world, nanotechnology has an endless potential in future to do coatings. Water-borne emulsion coatings included with nanoparticles to avoid microbial growth on metals and materials are emerging in the global coatings market.

The future innovative coating technology should not use hazardous materials for coatings, and mostly recycled materials will be cost effective if used as coatings. Recently, research has been focused on 'Green coatings' by incorporating the waste materials to avoid harmful chemicals, reduce carbon footprints, release volatile organic compounds, and reduce the waste. The future of green coating is booming as it is environmentally friendly and enhances the functionality of the surface.

Acknowledgment

The authors thank MHRD-SPARC (SPARC/2018-2019/P872/SL) for the funding.

References

Abdullah Dar, M. 2011. "A review: Plant extracts and oils as corrosion inhibitors in aggressive media." *Industrial Lubrication and* Tribology, 63(4): 227–233. Emerald Group Publishing Limited.

Buhri, Saleh Ali AL, Dinesh Keloth Kaithari, and Elansezhian Rasu. 2016. "Development of corrosion resistance coatings for sea water pipeline." *International Journal of Students' Research in Technology & Management* 4 (2): 24–9.

Caillol, Sylvain, Myriam Desroches, Gilles Boutevin, Cédric Loubat, Rémi Auvergne, and Bernard Boutevin. 2012. "Synthesis of new polyester polyols from epoxidized vegetable oils and biobased acids." *European Journal of Lipid Science and Technology* 114 (12). John Wiley & Sons, Ltd: 1447–59.

Cassar, L. 2005. "Nanotechnology and photocatalysis in cementitious materials." *Nicom'* 2 (13–16): 1–7.

Chaudhari, Ashok, Vikas Gite, Sandip Rajput, Pramod Mahulikar, and Ravindra Kulkarni. 2013. "Development of eco-friendly polyurethane coatings based on neem oil polyetheramide." *Industrial Crops and Products* 50 (October): 550–56.

Chuayjuljit, S., A. Maungchareon, and O. Saravari. 2010. "Preparation and properties of palm oil-based rigid polyurethane nanocomposite foams." *Journal of Reinforced Plastics and Composites* 29 (2). SAGE Publications: 218–25.

Dai, Honghai, Liting Yang, Bo Lin, Chengshuang Wang, and Guang Shi. 2009. "Synthesis and characterization of the different soy-based polyols by ring opening of epoxidized soybean oil with methanol, 1,2-ethanediol and 1,2-propanediol." *JAOCS, Journal of the American Oil Chemists' Society* 86 (3). Springer: 261–67.

Enderus, N. F., and S. M. Tahir. 2017. "Green waste cooking oil-based rigid poly-urethane foam." In *IOP Conference Series: Materials Science and Engineering*, 271:012062.

Erhan, Sevim Z. 2005. *Industrial Uses of Vegetable Oil. Industrial Uses of Vegetable Oil.* AOCS Publishing, New York.

Gite, V. V., R. D. Kulkarni, D. G. Hundiwale, and U. R. Kapadi. 2006. "Synthesis and characterisation of polyurethane coatings based on trimer of Isophorone Diisocyanate (IPDI) and monoglycerides of oils." *Surface Coatings International Part B: Coatings Transactions* 89 (2). Springer: 117–22.

Guo, Andrew, Youngjin Cho, and Zoran S. Petrović. 2000. "Structure and proper-ties of halogenated and nonhalogenated soy-based polyols." *Journal of Polymer Science, Part A: Polymer Chemistry* 38 (21). John Wiley & Sons Inc: 3900–10.

Guo, Liya, Weiyong Yuan, Zhisong Lu, and Chang Ming Li. 2013. "Polymer/nanosil-ver composite coatings for antibacterial applications." *Colloids and Surfaces A: Physicochemical and Engineering Aspects* 439 (December). Elsevier B.V.: 69–83.

Javni, Ivan, Zoran S. Petrović, Andrew Guo, and Rachel Fuller. 2000. "Thermal stabil-ity of polyurethanes based on vegetable oils." *Journal of Applied Polymer Science* 77 (8). John Wiley & Sons, Ltd: 1723–34.

Kiatsimkul, Pim pahn, Galen J. Suppes, Fu hung Hsieh, Zuleica Lozada, and Yuan Chan Tu. 2008. "Preparation of high hydroxyl equivalent weight polyols from vegetable oils." *Industrial Crops and Products* 27 (3). Elsevier: 257–64.

Lahhit, N., A. Bouyanzer, J. M. Desjobert, B. Hammouti, R. Salghi, J. Costa, C. Jama, F. Bentiss, and L. Majidi. 2011. "Fennel (Foeniculum Vulgare) essential oil as green corrosion inhibitor of carbon steel in hydrochloric acid solution." *Portugaliae Electrochimica Acta* 29 (2). Sociedade Portuguesa de Electroquimica: 127–38.

Li, Yebo, Xiaolan Luo, and Shengjun Hu. 2015. *Bio-Based Polyols and Polyurethanes*, Green Chemistry for Sustainability, 1:79, Heidelberg: Springer.

Li, Yebo, Xiaolan Luo, and Shengjun Hu. 2015. "Polyols and polyurethanes from protein-based feedstocks." 65–79, In: Bio-based Polyols and Polyurethanes. Springer Briefs in Molecular Science. Springer, Cham.

Miao, Shida, Songping Zhang, Zhiguo Su, and Ping Wang. 2010. "A novel vegetable oil-lactate hybrid monomer for synthesis of high-T_g polyurethanes." *Journal of Polymer Science Part A: Polymer Chemistry* 48 (1). John Wiley & Sons, Ltd: 243–50.

Mohd Tahir, Syuhada, Wan Norfirdaus Wan Salleh, Nur Syahamatun Nor Hadid, Nor Fatihah Enderus, and Nurul Aina Ismail. 2016. "Synthesis of waste cooking oil-based polyol using sugarcane bagasse activated carbon and transesterifica-tion reaction for rigid polyurethane foam." *Materials Science Forum*, 846. Trans Tech Publications Ltd.: 690–96.

Pan, Xiao, and Dean C. Webster. 2012. "New biobased high functionality polyols and their use in polyurethane coatings." *ChemSusChem* 5 (2). Wiley-VCH Verlag: 419–29.

Panadare, D C, and V K Rathod. 2015. "Applications of waste cooking oil other than biodiesel : A review." *Iranian Journal of Chemical Engineering* 12 (3): 55–76.

Patil, Chandrashekhar K., Sandip D. Rajput, Ravindra J. Marathe, Ravindra D. Kulkarni, Hemant Phadnis, Daewon Sohn, Pramod P. Mahulikar, and Vikas V. Gite. 2017. "Synthesis of bio-based polyurethane coatings from vegetable oil and dicarboxylic acids." *Progress in Organic Coatings* 106 (May). Elsevier B.V.: 87–95.

Raja, Pandian Bothi, and Mathur Gopalakrishnan Sethuraman. 2008. "Natural prod-ucts as corrosion inhibitor for metals in corrosive media - A review." *Materials Letters* 62 (1). North-Holland: 113–16.

Salimon, Jumat, Nadia Salih, and Emad Yousif. 2012. "Industrial development and applications of plant oils and their biobased oleochemicals." *Arabian Journal of Chemistry* 5(2): 135–145. Elsevier.

Sathiyanathan, R. A. L., S. Maruthamuthu, M. Selvanayagam, S. Mohanan, and N. Palaniswamy. 2005. "Corrosion inhibition of mild steel by ethanolic extracts of ricinus communis leaves." *Indian Journal of Chemical Technology* 12: 356–60.

Shukla, V., M. Singh, D. K. Singh, and Ri Shukla. 2005. "The role of castor oil in epoxy and polyamide systems for coating and adhesive application." *Surface Coatings International Part B: Coatings Transactions* 88 (3). Oil and Colour Chemists' Association: 217–20.

Srivastava, Kumkum, and Poonam Srivastava. 1981. "Studies-on plant materials as corrosion inhibitors." *British Corrosion Journal* 16 (4). Taylor & Francis: 221–23.

Stirna, U., U. Cabulis, and I. Beverte. 2008. "Water-blown polyisocyanurate foams from vegetable oil polyols." *Journal of Cellular Plastics* 44 (2): 139–60.

Thakur, Suman, and Niranjan Karak. 2013. "Castor oil-based hyperbranched polyurethanes as advanced surface coating materials." *Progress in Organic Coatings* 76 (1). Elsevier: 157–64.

Toker, R. D., N. Kayaman-Apohan, and M. V. Kahraman. 2013. "UV-curable nano-silver containing polyurethane based organic-inorganic hybrid coatings." *Progress in Organic Coatings* 76 (9). Elsevier: 1243–50.

Ullah, Zahoor, Mohamad Azmi Bustam, and Zakaria Man. 2014. "Characterization of waste palm cooking oil for biodiesel production." *International Journal of Chemical Engineering and Applications* 5 (2): 134–37.

Wang, Cheng-Shuang, Li-Ting Yang, Bao-Lian Ni, and Guang Shi. 2009. "Polyurethane networks from different soy-based polyols by the ring opening of epoxidized soybean oil with methanol, glycol, and 1,2-propanediol." *Journal of Applied Polymer Science* 114 (1). John Wiley & Sons, Ltd: 125–31.

Xia, Ying, and Richard C. Larock. 2010. "Vegetable oil-based polymeric materials: synthesis, properties, and applications." *Green Chemistry* 12 (11). The Royal Society of Chemistry: 1893–909.

Xu, Wentao, Chunfeng Ma, Jielin Ma, Tiansheng Gan, and Guangzhao Zhang. 2014. "Marine biofouling resistance of polyurethane with biodegradation and hydrolyzation." *ACS Applied Materials and Interfaces* 6 (6). American Chemical Society: 4017–24.

Index

Printed in the United States
by Baker & Taylor Publisher Services